AF541622

PRINCIPLES OF BIOINFORMATICS

By

Dr. Sharad Srivastava

Dept. of Zoology
Vikram University
Ujjain

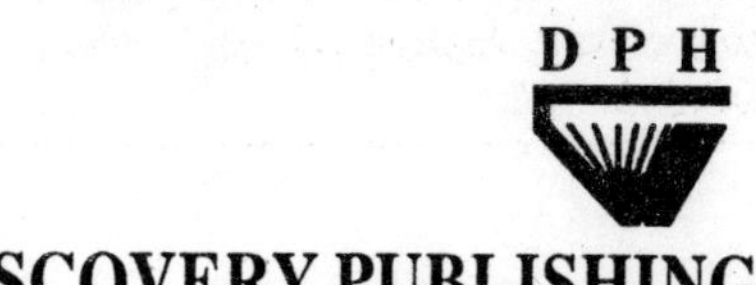

DISCOVERY PUBLISHING HOUSE PVT. LTD.
NEW DELHI-110 002

Published by:
Namit Wasan
DISCOVERY PUBLISHING HOUSE PVT. LTD.
4383/4B, Ansari Road, Darya Ganj
New Delhi-110 002 (India)
Phone : +91-11-23279245; 23253475; 43596065
E-mail : discoverybooksindia@gmail.com
discoverypublishinghouse@gmail.com
namitwasan9@gmail.com
web : www.discoverypublishinggroup.com

Edition: 2020

ISBN: 978-81-8356-355-0

Principles of Bioinformatics

Printed at:
Infinity Imaging Systems
Delhi

Preface

The present text offers a clear exposition of the *Principles of Bioinformatics*. Accessible to students in both biology and computer science, it strikes a unique balance between rigorous mathematics and practical techniques, emphasizing the ideas underlying computational rather than offering a collection of apparently unrelated problems.

The book introduces biological and computational ideas together, linking issues in computer science to biology and thus capturing the interest of students in both subjects. It demonstrates that relatively few design techniques can be used to solve a large number of practical problems in biology, and presents this material intuitively.

The present title is one of the first books on bioinformatics that can be used by students at an undergraduate level. It includes a dual table of contents, organized by biological idea and biological topic; discussions of biologically relevant problems, including a detailed problem formulation and one or more solutions for each. These interesting vignettes offer students a glimpse of the inspirations and motivations for real work in bioinformatics, making the concepts presented in the text more concrete and the techniques more approachable.

There can be no claim to originality except in the manner of treatment and much of the information has been obtained from the books and scientific journals available in the different libraries.

The author expresses his thanks to his friends and colleagues whose continue inspirations have initiated him to bring out this book.

The author is painfully aware of the shortcomings, errors and misprints that have crept in, and shall be grateful to receive suggestion for improvement of the next edition from all the readers.

The author expresses his gratitude to Mr. Wasan and staff of M/s Discovery Publishing House for their whole hearted co-operation in the publication of this book.

Author

Contents

1

INTRODUCTION

Bioinformatics represents a new, growing area of science that uses computational approaches to answer biological questions. Answering these questions requires that investigators take advantage of large, complex data sets (both public and private) in a rigorous fashion to reach valid, biological conclusions. The potential of such an approach is beginning to change the fundamental way in which basic science is done, helping to more efficiently guide experimental design in the laboratory.

With the explosion of sequence and structural information available to researchers, the field of bioinformatics is playing an increasingly large role in the study of fundamental biomedical problems. The challenge facing computational biologists will be to aid in gene discovery and in the design of molecular modeling, site-directed mutagenesis, and experiments of other types that can potentially reveal previously unknown relationships with respect to the structure and function of genes and proteins. This challenge becomes particularly daunting in light of the vast amount of data that has been produced by the Human Genome Project and other systematic sequencing efforts to date.

Before embarking on any practical discussion of computational methods in solving biological problems, it is necessary to lay the common groundwork that will enable users to both access and implement the algorithms and tools discussed in this book. We begin with a review of the Internet and its terminology, discussing major Internet protocol classes as well, without becoming overly engaged in the engineering minutiae underlying these protocols. A more in-depth treatment on the inner workings of these protocols may be found in a number of well-written reference books intended for the lay audience. This chapter will also discuss matters of connectivity, ranging from simple modem connections to digital subscriber lines (DSL). Finally, we will address one of the most common problems that has arisen with the proliferation of Web pages throughout the world—finding useful information on the World Wide Web.

INTERNET BASICS

Despite the impression that it is a single entity, the Internet is actually a network of networks, composed of interconnected local and regional networks in over 100 countries. Although work on remote communications began in the early 1960s, the true origins of the Internet lie with a research project on networking at the Advanced Research Projects Agency (ARPA) of the US Department of Defense in 1969 named ARPANET. The original ARPANET connected four nodes on the West Coast, with the immediate goal of being able to transmit information on defense-related research between laboratories. A number of different network projects subsequently surfaced, with the next landmark developments coming over 10 years later. In 1981, BITNET ("Because It's Time") was introduced, providing point-to-point connections between universities for the transfer of electronic mail and files. In 1982, ARPA introduced the Transmission Control Protocol (TCP) and the Internet Protocol (IP): TCP/IP allowed different networks to be connected to and communicate with one another, creating the system in place today. Most users, however, are content to leave the details of *how* the Internet works to their systems administrators; the relevant fact to most is that it *does* work.

Once the machines on a network have been connected to one another, there needs to be an unambiguous way to specify a single computer so that messages and files actually find their intended recipient. To accomplish this, all machines directly connected to the Internet have an *IP number*. IP addresses are unique, identifying one and only one machine. The IP address is made up of four numbers separated by periods; for example, the IP address for the main file server at the National Center for Biotechnology Information (NCBI) at the National Institutes of Health (NIH) is 130.14.25.1. The numbers themselves represent, from left to right, the domain (130.14 for NIH), the subnet (.25 for the National Library of Medicine at NIH), and the machine itself (.1). The use of IP numbers aids the computers in directing data: however, it is obviously very difficult for users to

Table 1.1. Top-level domain names

Top-level domain names	
.com	Commercial site
.edu	Educational site
.gov	Government site
.mil	Military site
.net	Gateway or network host
.org	Private (usually not-for-profit) organizations
Examples of top-level domain names used outside the United States	
.ca	Canadian site
.ac.uk	Academic site in the United Kingdom
.co.uk	Commercial site in the United Kingdom
Generic top-level domains proposed by IAHC	
.firm	Firms or businesses
.shop	Businesses offering goods to purchase (stores)
.web	Entities emphasizing activities relating to the World Wide Web
.arts	Cultural and entertainment organizations
.rec	Recreational organizations
.info	Information sources
.nom	Personal names (e.g., *yourlastname.nom*)

A complete listing of domain suffixes, including country codes, can be found at http://www.current.net/resources/directory/noframes/nf.domains.html.

remember these strings, so IP addresses often have associated with them a *fully qualified domain name* (FQDN) that is dynamically translated in the background by *domain name servers*. Going back to the NCBI example, rather than use 130.14.25.1 to access the NCBI computer, a user could instead use ncbi.nlm.nih.gov and achieve the same result. Reading from left to right, notice that the IP address goes from least to most specific, whereas the FQDN equivalent goes from most specific to least. The name of any given computer can then be thought of as taking the general form *computer.domain*, with the top-level domain (the portion coming after the last period in the FQDN) falling into one of the broad categories. Outside the United States, the top-level domain names *may* be replaced with a two-letter code specifying the country in which the machine is located (e.g., .ca for Canada and .uk for the United Kingdom). In an effort to anticipate the needs of Internet users in the future, as well as to try to erase the arbitrary line between top-level domain names based on country, the now-dissolved International Ad Hoc Committee (IAHC) was charged with developing a new framework of generic top-level domains (gTLD). The new, recommended gTLDs were set forth in a document entitled *The Generic Top Level Domain Memorandum of Understanding* (gTLD-MOU): these gTLDs are overseen by a number of governing bodies.

The most concrete measure of the size of the Internet lies in actually counting the number of machines physically connected to it. The Internet Software Consortium (ISC) conducts an Internet Domain Survey twice each year to count these machines, otherwise known as *hosts*. In performing this survey, ISC considers not only how many hostnames have been assigned, but how many of those are actually in use; a hostname might be issued, but the requestor may be holding the name in abeyance for future use. To test for this, a representative sample of host machines are sent a probe (a "ping"), with a signal being sent back to the originating machine if the host was indeed found. The rate of growth of the number of hosts has been phenomenal: from a paltry 213 hosts in August 1981, the Internet now has more than 60 million "live" hosts. The doubling time for the number of hosts is on the order of 18 months. At this time, most of this growth has come from the commercial sector, capitalizing on the growing popularity of multimedia platforms for advertising and communications such as the World Wide Web.

Connecting to the Internet

Of course, before being able to use all the resources that the Internet has to offer, one needs to actually make a physical connection between one's own computer and "the information superhighway." For purposes of this discussion, the elements of this connection have been separated into two discrete parts: the actual, physical connection (meaning the "wire" running from one's computer to the Internet backbone) and the service provider, who handles issues of routing and content once connected. Keep in mind that, in practice, these are not necessarily treated as two separate parts—for instance, one's service provider may also be the same company that will run cables or fibers right into one's home or office.

Copper Wires, Coaxial Cables, and Fiber Optics

Traditionally, users attempting to connect to the Internet away from the office had one and only one option—a modem, which uses the existing copper twisted-pair cables carrying telephone signals to transmit data. Data transfer rates using modems are relatively slow, allowing for data transmission in the range of 28.8 to 56 kilobits per second (kbps). The problem with using conventional copper wire to

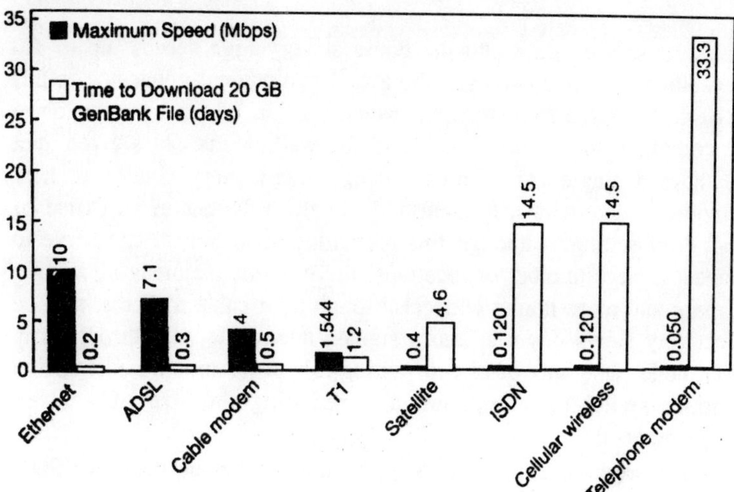

Fig. 1.1. Performance of various types of internet connections, by maximum throughput.

transmit data lies not in the copper wire itself but in the switches that are found along the way that route information to their intended destinations. These switches were designed for the efficient and effective transfer of voice data but were never intended to handle the high-speed transmission of data. Although most people still use modems from their home, a number of new technologies are already in place and will become more and more prevalent for accessing the Internet away from hardwired Ethernet networks.

The first of these "new solutions" is the integrated services digital network or ISDN. The advent of ISDN was originally heralded as the way to bring the Internet into the home in a speed-efficient manner; however, it required that special wiring be brought into the home. It also required that users be within a fixed distance from a central office, on the order of 20,000 feet or less. The cost of running this special, dedicated wiring, along with a per-minute pricing structure, effectively placed ISDN out of reach for most individuals. Although ISDN is still available in many areas, this type of service is quickly being supplanted by more cost-effective alternatives.

In looking at alternatives that did not require new wiring, cable television providers began to look at ways in which the coaxial cable already running into a substantial number of households could be used to also transmit data. Cable companies are able to use bandwidth that is not being used to transmit television signals (effectively, unused channels) to push data into the home at very high speeds, up to 4.0 megabits per second (Mbps). The actual computer is connected to this network through a cable modem, which uses an Ethernet connection to the computer and a coaxial cable to the wall. Homes in a given area all share a single cable, in a wiring scheme very similar to how individual computers are connected via the Ethernet in an office or laboratory setting. Although this branching arrangement can serve to connect a large number of locations, there is one major disadvantage: as more and more homes connect through their cable modems, service effectively slows down as more signals attempt to pass through any given node. One way of circumventing this problem is the installation of more switching equipment and reducing the size of a given "neighbourhood."

Because the local telephone companies were the primary ISDN providers, they quickly turned their attention to ways that the existing, conventional copper wire already in the home could be used to transmit data at high speed. The solution here is the digital subscriber line or DSL. By using new, dedicated switches that are designed for rapid data transfer, DSL providers can circumvent the old voice switches that slowed down transfer speeds. Depending on the user's distance from the central office and whether a particular neighbourhood has been wired for DSL service, speeds are on the order of 0.8 to 7.1 Mbps. The data transfers do not interfere with voice signals, and users can use the telephone while connected to the Internet; the signals are "split" by a special modem that passes the data signals to the computer and a microfilter that passes voice signals to the handset. There is a special type of DSL called *asynchronous* DSL or ADSL. This is the variety of DSL service that is becoming more and more prevalent. Most home users download much more information than they send out; therefore, systems are engineered to provide super-fast transmission in the "in" direction, with transmissions in the "out" direction being 5-10 times slower. Using this approach maximizes the amount of bandwidth that can be used without necessitating new wiring. One of the advantages of ADSL over cable is that ADSL subscribers effectively have a direct line to the central office, meaning that they do not have to compete with their neighbours for bandwidth. This, of course, comes at a price; at the time of this writing, ADSL connectivity options were on the order of twice as expensive as cable Internet, but this will vary from region to region.

Some of the newer technologies involve wireless connections to the Internet. These include using one's own cell phone or a special cell phone service (such as Ricochet) to upload and download information. These cellular providers can provide speeds on the order of 28.8-128 kbps, depending on the density of cellular towers in the service area. Fixed-point wireless services can be substantially faster because the cellular phone does not have to "find" the closest tower at any given time. Along these same lines, satellite providers are also coming on-line. These providers allow for data download directly to a satellite dish with a southern exposure, with uploads occurring through traditional telephone lines. Along the satellite option has the potential to be among the fastest of the options discussed, current operating speeds are only on the order of 400 kbps.

Content Providers vs. ISPs

Once an appropriately fast and price-effective connectivity solution is found, users will then need to actually connect to some sort of service that will enable them to traverse the Internet space. The two major categories in this respect are *online services* and *Internet service providers* (ISPs). Online services, such as America Online (AOL) and CompuServe, offer a large number of interactive digital services, including information retrieval, electronic mail (E-mail), bulletin boards, and "chat rooms," where users who are online at the same time can converse about any number of subjects. Although the online services now provide access to the World Wide Web, most of the specialized features and services available through these systems reside in a proprietary, closed network. Once a connection has been made between the user's computer and the online service, one can access the special features, or content, of these systems without ever leaving the online system's host computer. Specialized content can range from access to online travel reservation systems to encyclopedias that are constantly being updated—items that are not available to nonsubscribers to the particular online service.

Internet service providers take the opposite tack. Instead of focusing on providing content, the ISPs provide the tools necessary for users to send and receive E-mail, upload and download files, and navigate around the World Wide Web, finding information at remote locations. The major advantage of ISPs is connection speed: often the smaller providers offer faster connection speeds than can be had from the online services. Most ISPs charge a monthly fee for unlimited use.

The line between online services and ISPs has already begun to blur. For instance, AOL's now monthly flat-fee pricing structure in the United States allows users to obtain all the proprietary content found on AOL as well as all the Internet tools available through ISPs, often at the same cost as a simple ISP connection. The extensive AOL network puts access to AOL as close as a local phone call in most of the United States, providing access to E-mail no matter where the user is located, a feature small, local ISPs cannot match. Not to be outdone, many of the major national ISP providers now also provide content through the concept of *portals*. Portals are Web pages that can be customized to the needs of the individual user and that serve as a jumping-off point to other sources of news or entertainment on the Net. In addition, many national firms such as Mindspring are able to match AOL's ease of connectivity on the road, and both ISPs and online providers are becoming more and more generous in providing users the capacity to publish their own Web pages. Developments such as this, coupled with the move of local telephone and cable companies into providing Internet access through new, faster fiber optic networks, foretell major changes in how people will access the Net in the future, changes that should favour the end user in both price and performance.

Electronic Mail

Most people are introduced to the Internet through the use of electronic mail or *E-mail*. The use of E-mail has become practically indispensable in many settings because of its convenience as a medium for sending, receiving, and replying to messages. Its advantages are many:

1. It is much quicker than the postal service or "snail mail."
2. Messages tend to be much clearer and more to the point than is the case for typical telephone or face-to-face conversations.
3. Recipients have more flexibility in deciding whether a response needs to be sent immediately, relatively soon, or at all, giving individuals more control over workflow.
4. It provides a convenient method by which messages can be filed or stored.
5. There is little or no cost involved in sending an E-mail message.

These and other advantages have pushed E-mail to the forefront of interpersonal communication in both industry and the academic community: however, users should be aware of several major disadvantages. First is the issue of security. As mail travels toward its recipient, it may pass through a number of remote nodes, at any one of which the message may be intercepted and read by someone with high-level access, such as a systems administrator. Second is the issue of privacy. In industrial settings, E-mail is often considered to be an asset of the company for use in official communication only and, as such, is subject to monitoring by supervisors. The opposite is often true in academic, quasi-academic, or research settings; for example, the National Institutes of Health's policy encourages personal use of E-mail within the bounds of certain published guidelines. The key words here are "published guidelines": no matter what the setting, users of E-mail systems should always find out their organization's policy regarding appropriate use and confidentiality so that they may use the tool properly and effectively. An excellent, basic guide to the effective use of E-mail is recommended.

Sending E-Mail

E-mail addresses take the general form *user@computer.domain*, where *user* is the name of the individual user and *computer.domain* specifies the actual computer that the E-mail account is located on. Like a postal letter, an E-mail message is comprised of an *envelope* or *header*, showing the E-mail addresses of sender and recipient, a line indicating the subject of the E-mail, and information about how the E-mail message actually traveled from the sender to the recipient. The header is followed by the actual message, or body, analogous to what would go inside a postal envelope.

E-mail programs vary widely, depending on both the platform and the needs of the users. Most often, the characteristics of the local area network (LAN) dictate what types of mail programs can be used, and the decision is often left to systems administrators rather than individual users. Among the most widely used E-mail packages with a graphical user interface are Eudora for the Macintosh and both Netscape Messenger and Microsoft Exchange for the Mac, Windows, and UNIX platforms. Text-based E-mail programs, which are accessed by logging in to a UNIX-based account, include Elm and Pine.

Bulk E-Mail

As with postal mail, there has been an upsurge in "spam" or "junk E-mail," where companies compile bulk lists of E-mail addresses for use in commercial promotions. Because most of these lists are compiled from online registration forms and similar sources, the best defense for remaining off these bulk E-mail lists is to be selective as to whom E-mail addresses are provided. Most newsgroups keep their mailing lists confidential; if in doubt and if this is a concern, one should ask.

E-Mail Servers

Most often, E-mail is thought of a way to simply send messages, whether it be to one recipient or many. It is also possible to use E-mail as a mechanism for making predictions or retrieving records from biological databases. Users can send E-mail messages in a format defining the action to be performed to remote computers known as *servers*; the servers will then perform the desired operation and E-mail back the results. Although this method is not interactive (in that the user cannot adjust parameters or have control over the execution of the method in real time), it does place the responsibility for hardware maintenance and software upgrades on the individuals maintaining the server, allowing users to concentrate on their results instead of on programming. For most of these servers, sending the message *help* to the server E-mail address will result in a detailed set of instructions for using that server being returned, including ways in which queries need to be formatted.

Aliases and Newsgroups

In the example, the E-mail message is being sent to a single recipient. One of the strengths of E-mail is that a single piece of E-mail can be sent to a large number of people. The primary mechanism for doing this is through *aliases*; a user can define a group of people within their mail program and give the group a special name or alias. Instead of using individual E-mail addresses for all of the people in the group, the user can just send the E-mail to the alias name, and the mail program will handle broadcasting the message to each person in that group. Setting up alias names is a tremendous time-saver even for small groups; it also ensures that all members of a given group actually receive all E-mail messages intended for the group.

The second mechanism for broadcasting messages is through *newsgroups*. This model works slightly differently in that the list of E-mail addresses is compiled and maintained on a remote computer through subscriptions, much like magazine subscriptions. To participate in a newsgroup discussions, one first would have to subscribe to the newsgroup of interest. Depending on the newsgroup, this is done either by sending an E-mail to the host server or by visiting the host's Web site and using a form to subscribe. For example, the BIOSCI newsgroups are among the most highly trafficked, offering a forum for discussion or the exchange of ideas in a wide variety of biological subject areas. Information on how to subscribe to one of the constituent BIOSCI newsgroups is posted on the BIOSCI Web site. To actually participate in the discussion, one would simply send an E-mail lo the address corresponding to the group that you wish to reach. For example, to post messages to the computational biology newsgroup, mail would simply be addressed to comp-bio@net.bio.net, and, once that mail is sent, *everyone* subscribing to that newsgroup would receive (and have the opportunity to respond to) that message. The ease of reaching a large audience in such a simple fashion is both a blessing and a curse, so many newsgroups require that postings be reviewed by a moderator before they get disseminated to the individual subscribers to assure that the contents of the message are actually of interest to the readers.

It is also possible to participate in newsgroups without having each and every piece of E-mail flood into one's private mailbox. Instead, interested participants can use news-reading software, such as NewsWatcher for the Macintosh, which provides access to the individual messages making up a discussion. The major advantage is that the user can pick and choose which messages to read by scanning the subject lines; the remainder can be discarded by a single operation. NewsWatcher is an example of what is known as a *client-server application*, the client software (here, NewsWatcher) runs on a client computer (a Macintosh), which in turn interacts with a machine at a remote location (the server). Client-server architecture is interactive in nature, with a direct connection being made between the client and server machines.

Once NewsWatcher is started, the user is presented with a list of newsgroups available to them. This list will vary, depending on the user's location, as system administrators have the discretion to allow or to block certain groups at a given site. From the rear-most window in the figure, the user double-clicks on the newsgroup of interest (here, *bionet.genome.arabidopsis*), which spawns the window shown in the center. At the top of the center window is the current unread message count, and any message within the list can be read by double-clicking on that particular line. This, in turn, spawns the last window (in the foreground), which shows the actual message. If a user decides not to read any of the messages, or is done reading individual messages, the balance of the messages within the newsgroup (center) window can be deleted by first choosing Select All from the File menu and then selecting Mark Read from the News menu. Once the newsgroup window is closed, the unread message count is reset to zero. Every time NewsWatcher is restarted, it will automatically poll the news server for new messages that have been created since the last session. As with most of the tools that will be discussed in this chapter, news-reading capability is built into Web browsers such as Netscape Navigator and Microsoft Internet Explorer.

File Transfer Protocol

Despite the many advantages afforded by E-mail in transmitting messages, many users have no doubt experienced frustration in trying to transmit files, or *attachments*, along with an E-mail message. The mere fact that a file can be attached to an E-mail message and sent does not mean that the recipient will be able to detach, decode, and actually use the attached file. Although more cross-platform E-mail packages such as Microsoft Exchange are being developed, the use of different E-mail packages by people at different locations means that sending files via E-mail is not an effective, foolproof method, at least in the short term. One solution to this problem is through the use of a *file transfer protocol* or FTP. The workings of FTP are quite simple: a connection is made between a user's computer (the *client*) and a remote server, and that connection remains in place for the duration of the FTP session. File transfers are very fast, at rates on the order of 5-10 kilobytes per second, with speeds varying with the time of day, the distance between the client and server machines, and the overall traffic on the network.

In the ordinary case, making an FTP connection and transferring files requires that a user have an account on the remote server. However, there are many files and programs that are made freely available, and access to those files does not require having an account on each and every machine where these programs are stored. Instead, connections are made using a system called *anonymous FTP*. Under this system, the user connects to the remote machine and, instead of entering a username/password pair, types anonymous as the username and enters their E-mail address in place of a password. Providing one's E-mail address allows the server's system administrators to compile access statistics that may, in turn, be of use to those actually providing the public files or programs.

Although FTP actually occurs within the UNIX environment, Macintosh and PC users can use programs that rely on graphical user interfaces (GUI, pronounced "gooey") to navigate through the UNIX directories on the FTP server. Users need not have any knowledge of UNIX commands to download files, instead, they select from pop-up menus and point and click their way through the UNIX file structure. The most popular FTP program on the Macintosh platform for FTP sessions is Fetch. Notice that the Automatic radio button is selected, meaning that Fetch will determine the appropriate type of file transfer to perform. This may be manually overridden by selecting either Text or Binary, depending on the nature of the file being transferred. As a rule, text files should be transferred as Text, programs or executables as Binary, and graphic format files such as PICT and TIFF files as Raw Data.

World Wide Web

Although FTP is of tremendous use in the transfer of files from one computer to another, it does suffer from some limitations. When working with FTP, once a user enters a particular directory, they can only see the names of the directories or files. To actually view what is within the files, it is necessary to physically download the files onto one's own computer. This inherent drawback led to the development of a number of *distributed document delivery systems* (DDDS), interactive client-server applications that allowed information to be viewed without having to perform a download. The first generation of DDDS development led to programs like Gopher, which allowed plain text to be viewed directly through a client-server application. From this evolved the most widely known and widely used DDDS, namely, the World Wide Web. The Web is an outgrowth of research performed at the European Nuclear Research Council (CERN) in 1989 that was aimed at sharing research data between several locations. That work led to a medium through which text, images, sounds, and videos could be delivered to users on demand, anywhere in the world.

Navigation on the World Wide Web

Navigation on the Web does not require advance knowledge of the location of the information being sought. Instead, users can navigate by clicking on specific text, buttons, or pictures. These clickable items are collectively known as *hyperlinks*. Once one of these hyperlinks is clicked, the user is taken to another Web location, which could be at the same site or halfway around the world. Each document displayed on the Web is called a *Web page*, and all of the related Web pages on a particular server are collectively called a *Web site*. Navigation strictly through the use of hyperlinks has been nicknamed "Web surfing."

Users can take a more direct approach to finding information by entering a specific address. One of the strengths of the Web is that the programs used to view Web pages (appropriately termed *browsers*) can be used to visit FTP and Gopher sites as well, somewhat obviating the need for separate Gopher or FTP applications. As such, a unified naming convention was introduced to indicate to the browser program both the location of the remote site and, more importantly, the type of information at that remote location so that the browser could properly display the data. This standard-form address is known as a *uniform resource locator*, or URL, and takes the general form *protocol://computer.domain*, where *protocol* specifies the type of site and *computer.domain* specifies the location. The *http* used for the protocol in World Wide Web URLs stands for *hypertext transfer protocol*, the method used in transferring Web files from the host computer to the client.

Table 1.2. Uniform resource locator (URL) format for each type of transfer protocol

General form	*protocol://computer.domain*
FTP site	*ftp://ftp.ncbi.nlm.nih.gov*
Gopher site	*gopher://gopher.iubio.indiana.edu*
Web site	*http:/www.nhgri.nih.gov*

Browsers

Browsers, which are used to look at Web pages, arc client-server applications that connect to a remote site, download the requested information at that site, and display the information on a user's monitor, then disconnecting from the remote host. The information retrieved from the remote host is in a platform-independent format named *hypertext markup language* (HTML). HTML code is strictly text-based, and any associated graphics or sounds for that document exist as separate files in a common format. For example, images may be stored and transferred in GIF format, a proprietary format developed by CompuServe for the quick and efficient transfer of graphics; other formats, such as JPEG and BMP, may also be used. Because of this, a browser can display any Web page on any type of computer, whether it be a Macintosh, IBM compatible, or UNIX machine. The text is usually displayed first, with the remaining elements being placed on the page as they are downloaded. With minor exception, a given Web page will look the same when the same browser is used on any of the above platforms. The two major players in the area of browser software are Netscape, with their Communicator product, and Microsoft, with Internet Explorer. As with many other areas where multiple software products are available, the choice between Netscape and Internet Explorer comes down to one of personal preference. Whereas the computer literati will debate the fine points of difference between these two packages, for the average user, both packages perform equally well and offer the same types of features, adequately addressing the Web-browser needs of most users.

It is worth mentioning that, although the Web is by definition a visually-based medium, it is also possible to travel through Web space and view documents without the associated graphics. For users limited to line-by-line terminals, a browser called Lynx is available. Developed at the University of Kansas, Lynx allows users to use their keyboard arrow keys to highlight and select hyperlinks, using their return key the same way that Netscape and Internet Explorer users would click their mouse.

Internet vs. Intranet

The Web is normally thought of as a way to communicate with people at a distance, but the same infrastructure can be used to connect people within an organization. Such *intranets* provide an easily accessible repository of relevant information, capitalizing on the simplicity of the Web interface. They also provide another channel for broadcast or confidential communication within the organization. Having an intranet is of particular value when members of an organization are physically separated, whether in different buildings or different cities. Intranets are protected: that is, people who are not on the organization's network are prohibited from accessing the internal Web pages: additional protections through the use of passwords are also common.

Finding Information on the World Wide Web

Most people find information on the Web the old-fashioned, way: by word of mouth, either using lists such as those preceding the References in the chapters of this book or by simply following hyperlinks put in place by Web authors. Continuously clicking from page to page can be a highly ineffective, way of finding information, though, especially when the information sought is of a very focused nature. One way of finding interesting and relevant Web sites is to consult *virtual libraries*, which are curated lists of Web resources arranged by subject. Virtual libraries of special interest to biologists include the WWW Virtual Library, maintained by Keith Rohison at Harvard, and the EBI BioCatalog, based at the European Bioinformatics Institute.

It is also possible to directly search the Web by using *search engines*. A search engine is simply a specialized program that can perform full-text or keyword searches on databases that catalog Web

Table 1.3. Important internet resources.

Domain names	
gTLD-MOU	*http://www.gtld-mou.org*
Internet Software Consortium	*http:/www.isc.org*
Electronic mail and Newsgroups	
BIOSCI Newsgroups	*http://www.bio.net/docs/biosci.FAQ.html*
Eudora	*http://www.eudora.com*
Microsoft Exchange	*http://www.microsoft.com/exchange/*
NewsWatcher	*ftp://ftp.acns.nwu.edu/pub/newswatcher/*
File Transfer Protocol	
Fetch 3.0/Mac	*http://www.dartmouth.edu/pages/softdev/fetch.html*
LeechFTP/PC	*http://stud.fh-heibronn.de/jdebis/leechftp/*
Internet Access	
America Online	*http://www.aol.com*
AT&T	*http://www.att.con/worldnet*
Bell Atlantic	*http://www.verizon.net*
Bell Canada	*http://www.bell.ca*
CompuServe	*http://www.compuserve.com*
Ricochet	*http://www.ricochet.net*
Telus	*http://www.telus.net*
Worldcom	*http://www.worldcom.com*
Virtual Libraries	
EBI BioCatalog	*http://www.ebi.ac.uk/biocat/biocat.html*
Amos' WWW Links Page	*http://www.expasy.ch/alinks.html*
NAR Database Collection	*http://www.nar.oupjournals.org*
WWW Virtual Library	*http://mcb.harvard.edu/BioLinks.html*
World Wide Web Browsers	
Internet Explorer	*http://explorer.msn.com/home.htm*
Lynx	*ftp://ftp2.cc.ukans.edu/pub/lynx*
Netscape Navigator	*http://home.netscape.com*
World Wide Web Search Engines	
AltaVista	*http://www.altavista.com*
Excite	*http://www.excite.com*

Google	*http://www.google.com*
HotBot	*http://hotbot.lycos.com*
Infoseek	*http://infoseek.go.com*
Lycos	*http://www.lycos.com*
Northern Light	*http://www.northernlight.com*
World Wide Web Meta-Search Engines	
MetaCrawler	*http://www.metacrawler.com*
Savvy Search	*http://www.savysearch.com*

content. The result of a search is a hyperlinked list of Web sites fitting the search criteria from which the user can visit any or all of the found sites. However, the search engines use slightly different methods in compiling their databases. One variation is the attempt to capture most or all of the text of every Web page that the search engine is able to find and catalog ("Web crawling"). Another technique is to catalog only the title of each Web page rather than its entire text. A third is to consider words that must appear next to each other or only relatively close to one another. Because of these differences in search-engine algorithms, the results returned by issuing the same query to a number of different search engines can produce wildly different results. The other important feature is that most of the numbers are exceedingly large, reflecting the overall size of the World Wide Web. Unless a particular search engine ranks its results by relevance (e.g., by scoring words in a title higher than words in the body of the Web page), the results obtained may not be particularly useful. Also keep in mind that, depending on the indexing scheme that the search engine is using, the found pages may actually no longer exist, leading the user to the dreaded "404 Not Found" error.

Compounding this problem is the issue of coverage—the number of Web pages that any given search engine is actually able to survey and analyze. A comprehensive study by Lawrence and Giles (1998) indicates that the coverage provided by any of the search engines studied is both small and highly variable. For example, the HotBot engine produced 57.5% coverage of what was estimated to be the size of the "indexable Web," whereas Lycos had only 4.41% coverage, a full order of magnitude less than HotBot. The most important conclusion from this study was that the extent of coverage increased as the number of search engines was increased and the results from those individual searches were combined. Combining the results obtained from the six search engines examined in this study produced coverage approaching 100%.

To address this point, a new class of search engines called *meta-search engines* have been developed. These programs will take the user's query and poll anywhere from 5-10 of the "traditional" search engines. The meta-search engine will then collect the results, filter out duplicates, and return a single, annotated list to the user. One big advantage is that the meta-search engines take relevance statistics into account, returning much smaller lists of results. Although the hit list is substantially smaller, it is much more likely to contain sites that directly address the original query. Because the programs must poll a number of different search engines, searches conducted this way obviously take longer to perform, but the higher degree of confidence in the compiled results for a given query outweighs the extra few minutes (and sometimes only seconds) of search time. Reliable and easy-to-use meta-search engines include MetaCrawler and Savvy Search.

2

Sequence Assembly and Finishing Methods

That genome centers are able to produce DNA sequence data at very high speed using robotics and sequencing instruments should not be taken to imply that sequencing is always straightforward or routine. The nature of genomes and the limitations of mapping and sequencing techniques still make such projects challenging and provide scope for improved computing methods. To understand the requirements of sequence assembly software, it is necessary to know a little about genome sequences and sequencing techniques. After a general introduction to the task of sequencing and assembly and an illustration of the reality of routine sequencing operations, we describe how the Staden package can be employed in effectively assembling sequence data. Other packages will include their own, often similar, methods, to deal with the problems.

The large-scale organization and local composition of DNA affects the difficulty and complication of determining its sequence. Genomes of higher organisms vary considerably in their relative GC-to-AT content and in the number and types of repetitive elements they contain. For example, within its components of known utility, the human genome consists of genes for proteins which occur in single or few copies, multigene families scattered throughout the chromosomes, and gene clusters in a variety of arrangements, including multiple copies of genes for proteins and RNAs needed in high abundance. A large proportion of the remainder of the genome consists of various types of repetitive elements including LINEs and SINEs of which Alu sequences are the most widely known. Many other simple elements are also repeated, sometimes hundreds of times.

The Sanger dideoxy sequencing technique, which is employed by all the major public genome sequencing projects, uses DNA polymerase to synthesize a complementary copy of the target DNA. The inclusion of dideoxy nucleotides (which terminate the growing DNA chains) in the reaction mixture ensures that a proportion of all the fragments so produced stop after each base, hence creating a complete set of nested sequences. The polymerase extends, in the 5′ to 3′ direction, a short segment of DNA (known as a *primer*) that has to be annealed to the 5′ end of the target sequence to initiate the process. Either the primers or the dideoxy "terminators" are fluorescently labeled. The fragments are run through a sieving material such as a gel, which separates them according to their length. The fluorescence of the separated fragments is measured, and the resulting traces are analyzed to produce the base calls. The results of individual experiments are known as "gel readings" or "reads." If the sequence is 100 bases in length, this requires separating fragments having length differing by only 1%, and this effectively limits the size of sequence that can be obtained by a single experiment to around 1,000 bases. Most sequence runs obtain approximately 500 bases.

There are many potential sources of problems when preparing the DNA "template" for sequencing, performing the sequencing reactions, and loading and running the sequencing instrument; all of these factors influence the reliability of the data obtained. The characteristics of individual segments of DNA also influence the difficulty of obtaining reliable sequence, dictating whether special techniques need to be applied.

At the level of the individual gel reading, potential sources of problems include (1) "compressions," in which secondary structure in the DNA fragments causes them to move anomalously in the gel so that more than one size of fragment may migrate to the same position; (2) "stops," where the polymerase has a tendency to dissociate and hence, in dye primer chemistry, produce a large band on the gel; (3) regions of extreme composition, including high GC or GT composition and homopolymer regions; and (4) repetitive DNA. These problems all affect the accuracy and reliability of the data obtained from these regions. At the level of joining the readings together in the correct order, the various types of repeats listed above are the major difficulty (especially if the problem is compounded by poor quality data).

As mentioned above, the primer sequence (of around 20 bases) must be complementary to the sequence at the 5′ end of the target DNA, and the sequence from each experiment is limited to around 1,000 bases. These two factors have led to the widespread use of the so-called "shotgun" strategy of DNA sequencing in which the target DNA is randomly broken into overlapping fragments of around 2,000 bases and cloned into a "sequencing vector." This sidesteps the problem of having to know and synthesize the sequence at the 5′ end of each segment of the target DNA because they now all have a common sequence, namely, that just to the 5′ side of the cloning site in the vector. The use of cloning techniques also provides a way of producing sufficient pure quantities of the DNA. However, this strategy creates the problem of not knowing the order of the fragments obtained from each sequencing experiment. An alternative method of producing random samples from along the target DNA while retaining the ability of using the same primer for each experiment is to use transposons. One further and important experimental strategy is to determine the sequence from both ends of each of the cloned fragments. Together, these "forward" and "reverse" readings, known as "read pairs," give data from opposite strands of the DNA and provide information about the relative positions and orientations of the pairs of readings from the same fragment or template. The recent introduction of capillary electrophoresis instruments has made the production of forward and reverse readings more useful because this technology removes the problem of gel lanes being confused and, hence, with readings being assigned the wrong clone name (and subsequent incorrect assignment of read pairs).

Use of Base Call Accuracy Estimates or Confidence Values

When the idea of using numerical estimates of base calling was put forward, it was expected that these methods would be supplied as part of the instrument manufacturers' base-calling software. Instead, the first useable numerical values were produced by the program phred, which was devised by an academic group. This was a very important step forward and has had a major impact on genome projects.

Through the program phrap, phred-derived confidence values have been used to improve the quality of sequence assembly. Through the Staden Package, confidence values are also extensively used during the finishing stages of sequencing projects. Here, much of the tedious and time-consuming trace checking is obviated by the software; the confidence values are used to decide if human expertise is required to adjudicate between conflicting base calls. The result is that the majority of conflicts need never be brought to the user's attention, greatly reducing the time required to check and edit a contig. When the phred-style confidence values became available, the Staden program was adjusted to work on the decibel scale that was then defined. Green's group have written a contig editor known as consed, which has some similarities to the one in gap4 described later in this chapter.

Phred produces a confidence value that defines the probability that the base call is correct. The values were calculated by analyzing the traces from which the readings are derived: peak spacing over seven peaks, peak height ratios over seven and three peaks, and the number of peaks to the nearest uncalled base. The confidence value is given by the formula

$$\text{C-value} = -10 \times \log_{10} (\text{probability of error})$$

A confidence value of 10 corresponds to an error rate of 1/10, 20 to 1/100, 30 to 1/1000, and so on.

Another program that does base calling and produces confidence values is TraceTuner. Also, the program ATQA calculates confidence values and produces probabilities for insertions and deletions but does not recall the bases.

Requirements for Assembly Software

The task that sequence assembly software needs to accomplish is to infer the original sequence from the evidence of the readings and ideally to give, for each base, a probability that it is correct. It might be thought that this is simply a matter of comparing and aligning all the readings in a fully automated process, but this is not the case. The combination of limited reading length, reading errors, and repeats means that this ideal of automation is still some way off; therefore, interactive software tools are essential for helping to solve the many problems that confront sequencing staffs worldwide. As should be clear from the section above, there is ample opportunity for joining readings in the wrong order or wrong orientation, for missing out or duplicating regions, as well as for making minor base assignment errors, insertions, or deletions. Tools are required to check for these contingencies, sometimes in concert with extra experiments such as restriction enzyme digests. At present, the sequencing process is often talked of as consisting of two parts, namely, assembly and finishing, but in practice there is considerable overlap between the two. Assembly is the process of attempting to order and align the readings, and finishing is the task of checking and editing the assembled data. This includes performing new sequencing experiments to fill gaps or to cover segments where the data is poor and adjudicating between conflicting readings when editing.

Global Assembly

The global sequence assembly problem can be divided into three steps:

1. Find all possible overlaps between readings by comparing each one, in both orientations, to all the others;
2. from the list of overlaps, produce the best layout of the readings;
3. from the alignment of the readings in the final layout, derive a consensus sequence.

Step 1 is usually performed in two stages. First, a rapid comparison is performed to find all pairs of reading that share an exact match of, for example, 14 consecutive bases. Second, those that contain these matches are aligned using dynamic programming methods. The alignments that satisfy some preset criteria are "stored" in a graph, in which the vertices represent the readings and the edges represent the overlaps. Several different algorithms have been published that can analyze and prune these graphs to produce a consistent left-to-right ordering, orientation, and positioning for the readings. The resulting layout of the readings usually still requires multiple sequence alignment, as it is based only on individual pairwise alignments, each of which may conflict with others that they overlap. Once this has been done, a consensus can be derived. Elegant descriptions of the assembly problem and particular algorithmic solutions can be found in Kececioglu and Myers (1995) and Myers (1995).

Working programs usually include a number of important and effective extra methods. All the readings can be pre-screened to see if they contain the sequences of known repeats. Those that do can be set aside or treated in other special ways. For example, the segments containing repeat elements can be ignored during the search for an exact initial match but then used during the alignment phase. The layout can be checked and altered to be consistent with known read-pair data. The quality of the alignments can be scored by using the confidence values of the bases, and these scores can be used when the overlap graph is analyzed to produce the layout.

There are several widely used global assembly engines. Those that are currently available free of charge to the academic community include phrap, FAKII, CAP3, the TIGR Assembler and gap4. Those available commercially include Sequencher and DNASTAR.

After the global assembly engines have done their best with the initial shotgun data, the readings will be arranged into overlapping sets, and it is part of the job of the "finishing" process to complete the project by obtaining readings to fill the gaps. At the end of the project, there will be only one overlapping set, and it should cover the whole of the target sequence. Although its usage has now been expanded to include any set of overlapping clones, even those for which the sequence is unknown, the word "contig" was originally defined to mean a set of sequence readings that overlap. A consensus sequence can be calculated for each contig. New readings obtained for the finishing process can be compared against the consensus sequences. If the reading overlaps the end of one contig, it can be extended; if it overlaps two, these can be joined. This more limited assembly problem can be performed by the global engines or more quickly by simpler algorithms that use the consensus sequence.

File Formats

Raw sequence assembly data consist of traces, base calls, and confidence values, and most programs store these in SCF-format files. In addition to this, processing programs need data about how the readings were obtained—which sequencing vector they were cloned into, which primer was used, which template they were derived from, and which chemistry was used. A variety of methods are in use to manage this data. Many groups store their readings in FASTA-format files and thus have to use so-called "naming conventions" or "naming schemes" in which the entry name is used to encode data about readings. Others use relational databases, such as ABI's BioLIMS system. The Staden package uses Experiment files. The format is identical to EMBL/SWISS-PROT entries in that two-character record-type identifiers begin each line, but the record types are extended to include the necessary data about sequence readings. This simple text format is easily parsed. A set of C programs for reading and writing SCF and Experiment file formats is available on the Staden package Web site. Assemblies describe readings, chemistries, contigs, alignments and edits, and, for this, a "gap4 database" is used for each sequencing project.

The remainder of this chapter illustrates the use of the Staden package as it is applied to sequence assembly data, but readers should be aware that other packages include similar features. Because the examples here use a specific package, we have not gone into the details of how to perform particular tasks but instead give a flavour of the possible operations by describing some of the components of the individual programs.

Preparing Readings for Assembly

Data must be passed through several preassembly steps before being entered into the gap4 database, usually via a program called pregap4, which can operate on batches of traces.

Usually, trace files that are in proprietary format, such as those produced by ABI sequencers, are first converted to SCF files. Confidence values are added to the SCF file, and then its Experiment file is initialized with copies of the base calls and confidence values. The trace file is not needed again until the assembly is edited, and, because these programs can uncompress them on the fly, they are now compressed. The trace files are not altered but are kept as archival data so that it is always possible to check the original base calls and traces. Any changes to the data before assembly are made to the copy of the sequence in the Experiment file. It is recommended that no changes are made to the data until readings can be viewed aligned with others. The Experiment file is augmented to include data about how the readings were obtained—which sequencing vector they were cloned into, which primer was used, which template they were derived from, and which chemistry was used. This information can be obtained from a variety of sources. Next, the readings are analyzed to mark segments that are of low confidence, which can aid some assembly engines. With the use of the information in the Experiment file, the reading is searched for the presence of sequencing vectors at each end. A similar search is then done for cloning vectors (for example, for a BAC or cosmid vector). A final check is then performed for missed vectors or other possible contaminant sequences such as *E. coli* or yeast. All of these, searches write their results back into the Experiment file, ready for use by later programs in the pipeline. For the assembly stage, pregap4 can use phrap, CAP3, FAKII, or gap4; at the time of this writing, we believe that phrap is the most effective method. Only gap4 can read and write to gap4 databases; therefore, if an external assembly engine is used, the resulting assembly is copied into the gap4 database in the "enter assembly" step. When processing has finished, pregap4 will produce a report containing information from each module and the final list of "passed" and "failed" sequences. From this stage on, all changes are made to the copies of the data in

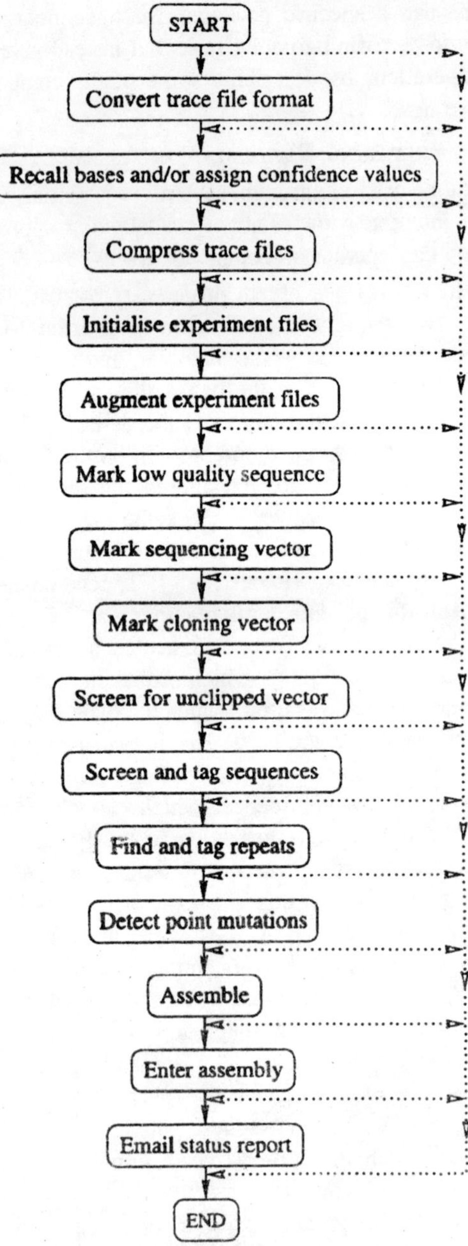

Fig. 2.1. A schematic of the tasks performed by the program pregap4 when processing a batch of sequence trace files.

the gap4 database and the Experiment files are no longer required. Pregap4 provides the interfaces to configure all of the above operations, to save the configuration for future use, and to perform the actual work.

Phrapview

Before other components of the gap4 package are described, a brief introduction to the phrapview program is warranted. Phrapview is distributed along with the phrap assembly engine and is a graphical viewer for phrap assemblies. It is intended to provide a "global" view of the assembly, complementing the individual base and trace view provided by consed. This global view focuses on information pertaining to possible incorrectness, incompleteness, or nonuniqueness of the phrap-generated assembly. Phrapview displays depth of coverage, forward-reverse read pairs, significant pairwise matches involving reads in different locations in the assembly, and chimeric reads.

The input to phrapview is a .view file, which is produced by running phrap with the View option. Note that phrapview does not perform any of the analyses itself; rather, it provides a way of displaying a file that contains an already completed analysis of the project. A screen dump for a typical phrapview display of a 40-kb cosmid sequencing project. In this display, colour-coded lines are drawn between read pairs, where red indicates a problem and black indicates "OK." Here, 48 problems and 120 OKs are reported. Read pairs will only be indicated properly if the read-naming convention assumed by phrap is used. The other types of data mentioned above can also be displayed in the same window.

Introduction to Gap4

Gap4 is an interactive program used for working on data from sequencing projects. It contains a comprehensive set of functions, many of which present their results graphically. Others, such as the Experiment Suggestion functions, produce textual output ready for parsing by external programs. One of its important components, used by many of the other functions, is the consensus algorithm. The gap4 database does not store the consensus sequence; rather, it is calculated whenever it is needed. When appropriate, it can be calculated separately for each strand, and, in the Contig Editor and Contig Joining Editor, it is instantly updated for each edit made. When phred-style confidence values are available, the algorithm uses them with strand and chemistry data to calculate a confidence value for each base in the consensus.

At the end of a project, the algorithm can produce a FASTA-format file or an Experiment file containing the consensus and its confidence values. Preprocessing programs used by pregap4 and routines within gap4 can add annotations to readings (for example, the position of an Alu segment or a custom primer). Throughout the text, these annotations are referred to as "tags."

The gap4 top-level window is divided into an Output Window for showing textual results, an Error Window for displaying error messages, and, at the top, an array of menus for selecting the main functions. The contents of the two text windows can be searched, edited, and saved. Each set of results is preceded by a header containing the time and date when it was generated. Most of gap4's tools are used within other windows, which are invoked when required or when an analytic function producing graphical output is selected. The most important windows are the Contig Selector, Contig Comparator, Template Display, Consistency Plot, Restriction Map, Contig Editor, Contig Joining Editor, and Trace Display. All of the graphical displays and the Contig Editors can be scrolled in register. The base of the graphical display windows usually contains an Information Line that shows short textual data about results or items touched by the mouse cursor. The next sections give short descriptions of the windows in order of their resolution.

Contig Selector

When an assembly database is opened, gap4 will bring up the Contig Selector, which is used to display, select and reorder contigs. In the Contig Selector, all contigs are shown as colinear horizontal lines separated by short vertical lines. The length of the horizontal lines is proportional to the length of the contigs, and their left-to-right order represents the current ordering of the contigs. The contig order can be changed by using the mouse to drag the lines representing the contigs. The Contig Selector can also be used to select contigs for processing. For example, clicking with the right mouse button on the line representing a contig will invoke a menu containing the commands that can be performed on that contig. As the mouse is moved over a contig, it is highlighted and the contig name and length are displayed in the Information Line.

In Contig Selector, along the top are three menus: File, View, and Results. Below this are buttons for zooming, activating the crosshair, and showing its position. The leftmost button, labeled Next, is used to sequentially step through sets of results. For example, if the user employs the Find Internal Joins option, which finds possible overlaps between contigs, the results are automatically sorted into descending order of overlap quality, and the Next button can be used to process them in that order. Each time the user clicks on the Next button, the Contig Joining Editor will be invoked to show the next overlap, enabling the user to examine the match and edit the two contigs and make the join. Below this row of buttons is the schematic of the contigs. The small boxes around the contig lines show the positions of tags on the readings and the consensus sequence.

Contig Comparator

Gap4 commands such as Find Internal Joins, Find Repeats, Check Assembly, and Find Read Pairs automatically transform the Contig Selector to produce the Contig Comparator. For this transformation, a copy of the Contig Selector is added at right angles to the original window to create a two-dimensional rectangular surface on which to display the results of comparing or checking contigs. It is therefore equivalent to the "dot plot" display commonly used for comparing pairs of sequences.

As mentioned above, Find Internal Joins compares the ends of contigs to see if there are possible overlaps. Find Repeats is similar, but, unlike Find Internal Joins, it does not require the found matches to continue to the ends of contigs. Check Assembly compares every reading with the segment of the consensus it overlaps to see how well they align. Those that align poorly are plotted along the main diagonal of the Contig Comparator. Find Read Pairs plots the positions of consistent read pairs that may indicate the order of contigs. Each of the functions plots its results as diagonal lines of different colours. Lines parallel to the main diagonal represent contigs that are in the correct orientation relative to one another. Those perpendicular to the main diagonal show results for which one contig would need to be reversed before the pair could be joined. The manual contig-dragging procedure mentioned above can be used to change the relative positions of contigs. As the contigs are dragged, the plotted results will automatically be moved to their corresponding new positions. Because this plot can simultaneously show the results of independent types of search, users are able to determine whether different analyses produce corroborating or conflicting evidence for the ordering of readings or contigs. The plotted results can be used to invoke a subset of commands by the use of pop-up menus. For example, if the user clicks the right mouse button over a result from Find Internal Joins, a menu containing

Invoke Join Editor and Invoke Contig Editors will pop up. If the user selects Invoke Join Editor, the Join Editor will be started with the two contigs aligned at the match position contained in the result. If required, one of the contigs will be complemented to allow their alignment.

Although they cannot be distinguished without their usual colour coding, this screen dump includes results for Find Internal Joins (black). Check Assembly (green), and Find Read Pairs (blue). Superimposed on the bottom left corner are the View menu and the Results. Manager menu. The crosshairs show the positions for a pair of contigs. The vertical line continues into the Contig Selector part of the display, and the position represented by the horizontal line is also duplicated there.

Template Display

The next level of resolution is the Template Display, which can show schematic plots of readings, templates, tags, restriction enzyme sites, and the consensus quality. It can show one or several contigs and uses colour coding to distinguish reading, primer, and template types. It is often used to view the results of the Order Contigs option, which uses read-pair data to find the most likely left-to-right order of the contigs. These data may contain errors due to misnaming of readings (and hence which templates they were obtained from); thus, it is useful to view the results and, if required, modify the order. Again, this can be done by dragging the lines representing the contigs (but this time in the Template Display rather than in the Contig Selector). If the contigs are dragged, the plot is immediately redrawn to reflect the new ordering.

Under the menus and buttons, the largest section of the display contains stacks of lines and arrows representing the readings and templates for the contigs (shown by single nonoverlapping lines surrounded by tags directly below). Beneath this is a representation of the consensus quality for one contig and below that its restriction map. Note that the high depth of coverage seen here is template coverage. An alternative plot in which only the readings (and not their templates) are drawn would show much lower coverage. The template and reading sections of the display are in two parts. The top part contains the templates that have been sequenced from both ends but that are in some way inconsistent—for example, given the current relative positions of their readings, they may have a length that is larger or greater than that expected or the two readings may point in opposite directions. In this screen dump, there are four inconsistent templates, three within contigs and one spanning a pair of contigs. Colour coding is used to distinguish between different types of inconsistencies and whether the inconsistencies involve readings within or between contigs. The rest of the data (mostly for templates sequenced from only one end) is plotted below the data for the inconsistent templates. Templates with only one reading are shown in dark blue. Superimposed on the lines representing the templates are forward readings, shown as light blue arrows, and reverse readings, shown as orange arrows. Templates in bright yellow have been sequenced from both ends, are consistent, and span a pair of contigs (and thus show the relative orientation and separation of the contigs). All data are scaled: the templates with one reading are given their expected length (here, 2,000 bases), templates within single contigs and with readings from both ends are shown with their actual length, and templates spanning contigs are shown with a length calculated from the positions of their readings. The screen dump contains two vertical lines that extend the full height of the plot. One is the crosshair, and the other shows the position of the Contig Editor editing cursor. Not only does this vertical line show the editing cursor position but, by using the mouse, it can be used to scroll the editor.

In the plot below the line representing the longest contig, the quality of its consensus is illustrated by colour and line height. Here, the "quality" of the consensus is calculated using a simple algorithm and is mostly useful for data that has no confidence values. If a consensus base in a contig has good data from both strands and the consensus, when calculated separately for each of the two strands, is in agreement, then the plot for that position will show nothing. However, if the two strands disagree, or there are good data for only one strand, lines will be plotted of a colour and height that signifies the problem. An alternative plot for data with confidence values, which shows the confidence for the consensus, is also available.

The position of a crosshair is shown in the two leftmost boxes in the top right-hand corner. The leftmost shows the distance in bases between the crosshair and the start of the contig underneath the crosshair. The middle box shows the distance between the crosshair and the start of the first contig. The bottom plot is of the positions of restriction enzyme cut sites, and the rightmost box at the top of the display shows the distance between two selected cut sites. Comparing the predicted restriction pattern and that obtained from the original physical map clone can be an important final check on the overall correctness of the assembly.

Consistency Display

An alternative way of viewing the data at a scale similar to that of the Template Display is provided by the Consistency Display, which can plot histograms of the reading coverage and the read-pair coverage. The blocky histogram shows two lines (one red, one black) depicting the number of readings covering each position in a contig. The coverage for each strand is summarized below by the two sets of horizontal lines; where a line appears, there is no data for the strand. In the example, there are no data for the minus strand at the left end of the contig and none for the plus strand just left of the right-hand cursor or crosshair.

Contig Editor

The most detailed level of resolution provided by gap4 is the Contig Editor and its associated Trace Display, which are used for the final checking and editing of the aligned readings. It makes this an efficient and rapid procedure by providing a variety of search methods and by automatically displaying the trace data for any problems found. Users can alter bases or their confidence values to resolve errors, and the consensus sequence is instantly updated. Editing should be performed only to obtain the correct consensus at the required level of confidence. The less editing done, the quicker the project will be completed, and the more original base calls visible in the Contig Editor, the easier it will be to check if any doubts arise in the future. The consensus calculation is the same as that used to produce the final consensus to be sent to the sequence library, and it will remove all padding characters that appear in the Contig Editor consensus. Note that the numbering shown along the top of the Contig Editor window can be set in two ways: positions can include or exclude consensus padding characters.

The Search dialogue is obscuring the right-hand side of the Editor window. The main components of the editor are the controls at the top, reading names on the left, sequences to their right, and status lines at the bottom. To the left of each reading name is the reading number, which is negative for readings that have been reversed and complemented. The reliability of the individual base calls is shown using gray scale—the lighter the background, the higher the confidence. Below the reading data is their consensus sequence, with its reliability again shown using gray scale. The first of the status lines, labeled "Strands," shows a summary of strand coverage. The left end of the segment of sequence being displayed is covered only by readings from one strand of the DNA (+), but the rest contains data from both strands (=). Two positions are marked with an exclamation mark (!), indicating that, when the consensus is calculated separately for each strand, they disagree.

Along the top of the editor window is a row of command buttons and menus. The rightmost pair of buttons provide help and exit. To their left are two menus, Settings and Commands. To the left of this is a button that initially displays the search dialogue: by pressing it again, the selected search will be performed. Further left is the Undo button; each time the user clicks on this box, the program reverses the previous edit command. The, next button, labeled Cutoffs, is used to toggle between showing or hiding the reading data of poor quality or are vector sequences. The next button to the left is the Edit Modes menu, which allows users to select which editing commands are enabled. The next command toggles between insert and replace and thus governs the effect of typing in the edit window. The Information Line at the bottom of the window can show information about readings, annotations, and base calls, depending on what is under the mouse cursor.

With the use of the Settings menu, the Contig Editor can be configured to display disagreements and edits. Disagreements between the consensus and individual base calls are shown in dark green. Edits are shown with a light green background, replacements/insertions are shown in pink, deletions are shown in red, and confidence value changes are shown in purple. As can be seen from the contents of the Search dialogue superimposed on the Contig Editor, there are many types of search that can be performed. For example, the cursor can explicitly be moved to any position in the contig, either ignoring or counting padding characters. Other operations include finding the next non-ACGT character in the consensus; the next edit, the next place where the confidence of the consensus falls below some user-defined value, and the next place where an edit has been performed, but there is no evidence for the change in the original data (i.e., the current base does not appear in any of the original readings covering the position). The type of search used varies depending on the user's role. For example, a user finishing a sequence project may search mainly for low-confidence consensus sequences, whereas someone in a supervisory role may go through and check every edit made.

The Contig Editor can display the traces for any reading or set of readings. The number of rows and columns of traces displayed can be set by the user. The traces scroll in register with one another, as well as with the cursor in the Contig Editor. Conversely, the Contig Editor cursor can be scrolled by the trace cursor. A typical view containing three traces is shown in the screen dump of the Contig Editor. These are the best two traces from each strand plus a trace from a reading that contains a disagreement with the consensus. With the use of the Settings menu, the program can be configured to automatically bring up this combination of traces for each problem located by the Next Search option. The histogram or vertical bars plotted top-down show the original confidence values for each base call. The reading number, together with the direction of the reading (+ or -) and the chemistry by which it was determined, is given at the top left of each subwindow. There are three buttons (Info, Diff, and Quit) arranged vertically, with X and Y scale bars to their right. The Info button produces a window containing notes about the trace. The Diff button is mostly used for mutation detection and causes a pair of traces to be subtracted from one another and the result plotted, hence revealing their differences.

Contig Join Editor

The Join Editor looks like a pair of Contig Editors stacked one above the other with a strip in between, labeled Diffs for displaying the disagreements between the two consensus sequences. The other two main differences are the Align and Lock buttons. The former performs an alignment between the two consensus sequences and inserts padding characters accordingly, and the latter toggles between scrolling the contigs separately or in unison. Each contig can be edited separately, and it is important that, before a join is confirmed, the alignment is checked over its full extent. The Contig Join Editor is usually invoked by clicking on a Find Internal Joins or Find Repeals result in the Contig Comparator, in which case the two contigs will appear aligned at the match found by these searches. This figure shows the right-hand end of one contig in the lower editor and the left-hand end of another in the top editor. The Cutoff or Hidden data are displayed for the right-hand contig but not for the left-hand contig.

Disassembling Readings

Pregap4 and gap4 use lists of readings and contigs for processing batches of data. Obviously, the input to pregap4 is usually a list of trace files. Within gap4, users can create these lists in a number of ways. For example, when using the Contig Editor, users can create a list of readings to be disassembled. These are automatically sent to the gap4 Disassemble Readings function when the editor is exited. Readings can either be removed from the database or from the contigs they were in. If they are removed only from the contigs, they each start new contigs of their own: therefore, if the Find Internal Joins function is applied to all the contigs in the database, it will reveal any overlaps they may have elsewhere. If they are removed from the database, the list can be used by the assembly function to reassemble them. During its disassembly, if a reading is the only one covering part of a contig, the contig will be automatically broken in two. Gap4 also has a specific function for this purpose that will break a contig at the start of a given reading.

Experiment Suggestion and Automation

Much of the foregoing has been to illustrate the types of tools that are available for solving difficult problems interactively, but ideally one would prefer to automate as much of the work as possible. Genome sequencing center staff have put many hours into developing their own in-house procedures for automating their work. The current release of gap4 enables this approach in two ways. The first is by use of its Experiment Suggestion functions, and the second is by use of the gap4 scripting language. Using its consensus algorithms, gap4 can analyze assembly databases to find segments of sequence that require further readings, either to resolve disagreements, to fill the requirement for data from both strands of the DNA, or to extend contigs to try join them to others. It can then suggest the experiments to perform and the templates to use because it has all the necessary information in its assembly database. However, in the current version of gap4, the types of experiments are limited, and it has no knowledge of the very latest techniques or the ones that may be available in any particular laboratory. For large laboratories using gap4, at present, it is better to use its algorithms to analyze the database and to use the results of that analysis with a set of external routines customized to local methods. As outlined below, this is enabled by the design of gap4.

An entirely new set of gap4 finishing functions is currently being tested at the Sanger Center. In addition to the consensus algorithm, these new functions employ an analysis of reading and template depth as well as a knowledge of sequencing chemistry to design sets of potential problem-solving experiments. These experiments are geared towards satisfying the finishing criteria specified by the user.

Table 2.1. Important internet resources

ATQA	*http://www.wagner.com*
CAP3	*http://www.cs.mtu.edu.faculty/Huang.html*
Consed	*http://bozeman.genome.washington.edu/consed/consed.html*
DNASTAR	*http://www.dnastar.com*
FAKII	*http://bozeman.genome.washington.edu/phrap.docs/phrap.html*
Phred	*http://bonzeman.genome.washington.edu/phrap.docs/phred.html*
Gap4	*http://mrc-lmb.com.ac.uk/pubseq/index.html*
Sequencher	*http://www.genecodes.com*
Staden package	*http://www.mrc-lmb.cam.ac.uk/pubseq/index.html*
TIGR Assembler	*http://www.tigr.org/softlab*
TraceTuner	*http://www.paracel.com/html/tracetuner.html*

Tcl is a portable and extensible scripting language written in the C language. The programs in the Staden package are written in the form of additional commands understood by the Tcl interpreter. Each of gap4's algorithms can therefore be used in Tcl scripts. This means that users can write scripts to perform analysis of gap4 databases (e.g., apply a consensus calculation or read-pair analysis algorithm), and these scripts could also be linked to local laboratory information management systems and robotics. A further type of automation is furnished by the fact that the gap4 Contig Editor can be driven by an external file of commands.

3

Molecular Biology Primer

To understand bioinformatics in any meaningful way, it is necessary for a computer scientist to understand some basic biology, just as it is necessary for a biologist to understand some basic computer science. This chapter provides a short and informal introduction to those biological fundamentals. We scanned existing bioinformatics books to find out how much biological material was "relevant" to those books and we were surprised how little biological knowledge was actually presented. It would be safe to say that the minimum biological background one needs in order to digest a typical bioinformatics book could fit into ten pages. In this chapter we give a brief introduction to biology that covers most of the computational concepts discussed in bioinformatics books. Some of the sections in this chapter are not directly related to the rest of the book, but we present them to convey the fascinating story of molecular biology in the twentieth century.

What is Life Made of?

Biology at the microscopic level began in 1665 when maverick and virtuoso performer of public animal dissections, Robert Hooke, discovered that organisms are composed of individual compartments called *cells*. Cell theory, further advanced by Matthias Schleiden and Theodor Schwann in the 1830s, marked an important milestone: it turned biology into a science beyond the reach of the naked eye. In many ways, the study of life became the study of cells.

A great diversity of cells exist in nature, but they all have some common features. All cells have a life cycle: they are born, eat, replicate, and die. During the life cycle, a cell has to make many important decisions. For example, if a cell were to attempt to replicate before it had collected all of the necessary nutrients to do so, the result would be a disaster. However, cells do not have brains. Instead, these decisions are manifested in complex networks of chemical reactions, called *pathways*, that synthesize new materials, break other materials down for spare parts, or signal that the time has come to eat or die. The amazingly reliable and complex algorithm that controls the life of the cell is still beyond our comprehension.

One can envision a cell as a complex mechanical system with many moving parts. Not only does it store all of the *information* necessary to make a complete replica of itself, it also contains all the *machinery* required to collect and manufacture its components, carry out the copying process, and kick-start its new offspring. In macroscopic terms, a cell would be roughly analogous to a car factory that could mine for ore, fabricate girders and concrete pillars, and assemble an exact working copy of itself, all the while building family sedans with no human intervention.

Despite the complexity of a cell, there seems to be a few organizing principles that are conserved across all organisms. All life on this planet depends on three types of molecule: DNA, RNA, and proteins. Roughly speaking, a cell's DNA holds a vast library describing how the cell works. RNA acts to transfer certain short pieces of this library to different places in the cell, at which point those smaller volumes of information are used as templates to synthesize proteins. Proteins form enzymes that perform biochemical reactions, send signals to other cells, form the body's major components (like the keratin in our skin), and otherwise perform the actual work of the cell. DNA, RNA, and proteins are examples of *strings* written in either the four-letter alphabet of DNA and RNA or the twenty-letter alphabet of proteins. This meshes well with Schrodinger's visionary idea about an "instruction book" of life scribbled m a secret code. It took a long time to figure out that DNA, RNA, and proteins are the main players in the cells. Below we give a brief summary of how this was discovered.

What is the Genetic Material?

Schleiden's and Schwann's studies of cells were further advanced by the discovery of threadlike chromosomes in the cell nucleii. Different organisms have different numbers of chromosomes, suggesting that they might carry information specific for each species. This fit well with the work of the Augustinian monk Gregor Mendel in the 1860s, whose experiments with garden peas suggested the existence of *genes* that were responsible for inheritance. Evidence that traits (more precisely, genes) are located on chromosomes came in the 1920s through the work of Thomas Morgan. Unlike Mendel, Morgan worked in New York City and lacked the garden space to cultivate peas, so he instead used fruit flies for his experiments: they have a short life span and produce numerous offspring. One of these offspring turned out to have white eyes, whereas wild flies had red eyes. This one white-eyed male fly born in Morgan's "fly room" in New York City became the cornerstone of modern genetics.

The white-eyed male fly was mated with its red-eyed sisters and the offspring were followed closely for a few generations. The analysis of offspring revealed that white eyes appeared predominantly in males, suggesting that a gene for eye colour resides on the X chromosome (which partly determines the gender of a fruit fly). Thus, Morgan suspected that genes were located on chromosomes. Of course, Morgan had no idea what chromosomes were themselves made of.

Morgan and his students proceeded to identify other mutations in flies and used ever more sophisticated techniques to assign these mutations to certain locations on chromosomes. Morgan postulated that the genes somehow responsible for these mutations were also positioned at these locations. His group showed that certain genes are inherited together, as if they were a single unit. For example, Morgan identified mutants with a black body colour (normal flies are gray) and mutants with vestigial wings. He proceeded to cross black flies with vestigial wings with gray flies with normal wings, expecting to see a number of gray flies with vestigial wings, gray flies with normal wings, black flies with vestigial wings, and black flies with normal wings. However, the experiment produced a surprisingly large number of normal flies (gray body, normal wings) and a surprisingly large number of double mutants (black body, vestigial wings). Morgan immediately proposed a hypothesis that such *linked* genes reside close together on a chromosome. Moreover, he theorized, the more tightly two genes are linked (i.e., the more often they are inherited together), the closer they are on a chromosome.

Morgan's student Alfred Sturtevant pursued Morgan's chromosome theory and constructed the first genetic map of a chromosome that showed the order of genes. Sturtevant studied three genes: *cn*, which determines eye colour; *b*, which determines body colour; and *vg*, which determines wing size. Sturtevant crossed double-mutant *b* and *vg* flies with normal flies and saw that about 17% of the offspring had only a single mutation. However, when Sturtevant crossed double-mutant *b* and *cn* flies he found that 9% of the offspring had only a single mutation. This implied that *b* and *cn* reside closer together than *b* and *vg*; a further experiment with *cn* and *vg* mutants demonstrated an 8% single mutation rate. Combined together, these three observations showed that *b* lies on one side of *cn* and *vg* on the other. By studying many genes in this way, it is possible to determine the ordering of genes. However, the nature of genes remained an elusive and abstract concept for many years, since it was not clear how genes encoded information and how they passed that information to the organism's progeny.

What do Genes do?

By the early 1940s, biologists understood that a cell's traits were inherent in its genetic information, that the genetic information was passed to its offspring, and that the genetic information was organized into genes that resided on chromosomes. They did not know what the chromosomes were made of or what the genes actually did to give rise to a cell's traits. George Beadle and Edward Taturn were the first to identify the job of the gene, without actually revealing the true nature of genetic information. They worked with the bread mold *Neurospora*, which can survive by consuming very simple nutrients like sucrose and salt. To be able to live on such a limited diet, *Neurospora* must have some proteins (enzymes) that are able to convert these simple nutrients into "real food" like amino acids and the other molecules necessary for life. It was known that proteins performed this type of chemical "work" in the cell.

In 1941 Beadle and Tatum irradiated *Neurospora* with x-rays and examined its growth on the usual "Spartan" medium. Not surprisingly, some irradiated *Neurospora* spores failed to grow on this diet. Beadle and Tatum conjectured that x-rays introduced some mutations that possibly "destroyed" one of the genes responsible for processing *Neurospora's* diet into real food. Which particular gene was destroyed remained unclear, but one of the experiments revealed that the irradiated *Neurospora* survived and even flourished when Beadle and Tatum supplemented its spartan diet with vitamin B_6. An immediate conclusion was that x-rays damaged a gene that produces a protein (enzyme) responsible for the synthesis of B_6. The simplest explanation for this observation was that the role of a gene was to produce proteins. "The rule of "one gene, one protein" remained the dominant thinking for the next half-century until biologists learned that one gene may produce a multitude of proteins.

What Molecule Codes for Genes?

DNA was discovered in 1869 by Johann Friedrich Miescher when he isolated a substance he called "nuclein" from the nuclei of white blood cells. By the early 1900s it was known that DNA (nuclein) was a long molecule consisting of four types of bases: adenine (A), thymine (T), guanine (G), and cytosine (C). Originally, biologists discovered five types of bases, the fifth being uracil (U), which is chemically similar to thymine. By the 1920s, nucleic acids were grouped into two classes called DNA and RNA, that differ slightly in their base composition: DNA uses T while RNA uses U.

DNA, or *deoxyribonucleic acid*, is a simple molecule consisting of a sugar (a common type of organic compound), a phosphate group (containing the element phosphorus), and one of four nitrogenous bases (A, T, G, or C). The chemical bonds linking together nucleotides in DNA are always the same such that the backbone of a DNA molecule is very regular. It is the A, T, C, and G bases that give "individuality" to each DNA molecule.

Ironically, for a long time biologists paid little attention to DNA since it was thought to be a repetitive molecule incapable of encoding genetic information. They thought that each nucleotide in DNA followed another in an unchanging long pattern like ATGCATGCATGCATG CATGC, like synthetic polymers. Such a simple sequence could not serve as Schrodinger's code-script, so biologists remained largely uninterested in DNA. This changed in 1944 when Oswald Avery and colleagues proved that genes indeed reside on DNA.

What is the Structure of DNA?

The modern DNA era began in 1953 when James Watson and Francis Crick determined the double helical structure of a DNA molecule. Just 3 years earlier, Erwin Chargaff discovered a surprising one-to-one ratio of the adenine-to-thymine and guanine-to-cytosine content in DNA (known as the *Chargaff rule*). In 1951, Maurice Wilkins and Rosalind Franklin obtained sharp x-ray images of DNA that suggested that DNA is a helical molecule.

Watson and Crick were facing a three-dimensional jigsaw puzzle: find a helical structure made out of DNA subunits that explains the Chargaff rule. When they learned of the Chargaff rule, Watson and Crick wondered whether A might be chemically attracted to T (and G to C) during DNA replication. If this was the case, then the "parental" strand of DNA would be complementary to the "child" strand, in the sense that ATGACC is complementary to TACTGG. After manipulating paper and metal Tinkertoy representations of bases Watson and Crick arrived at the very simple and elegant double-stranded helical structure of DNA. The two strands were held together by hydrogen bonds between specific base pairings: A-T and C-G. The key ingredient in their discovery was the chemical logic begind the complementary relationship between nucleotides in each strand—it explained the Chargaff rule, since A was predicted to pair with T, and C with G. Thus, the nucleotide string of one strand completely defined the nucleotide string of the other. This is, in fact, the key to DNA replication, and the missing link between the DNA molecule and heredity. As Watson and Crick gently put it in their one-page paper on April 25, 1953: "It has not escaped our notice that the specific pairing we have postulated immediately suggests a possible copying mechanism for the genetic material."

What Carries Information between DNA and Proteins?

The double helix provided the key to DNA replication, but the question remained as to how DNA (a long but simple molecule) generates an enormous variety of different proteins. The DNA content of a cell does not change over time, but the concentrations of different proteins do. DNA is written in a four-letter alphabet while proteins are written in a twenty-letter alphabet. The key insight was that different pieces of a long DNA molecule coded for different proteins. But what was the code that translated texts written in a four-letter alphabet into texts written in a twenty-letter alphabet? How was this code read and executed?

First, we must realize that there are two types of cells: those that encapsulate their DNA in a *nucleus* and those that do not. The former are referred to as *eukaryotic* cells and the latter are *prokaryotic* cells. All multicellular organisms (like flies or humans) are eukaryotic, while most unicellular organisms (like bacteria) are prokaryotic. For our purposes, the major difference between prokaryotes and eukaryotes is that prokaryotic genes are continuous strings, while they are broken into pieces (called *exons*) in eukaryotes. Human genes may be broken into as many as 50 exons, separated by seemingly meaningless pieces called *introns*, whose function researchers are still trying to determine.

Understanding the connection between DNA and proteins began with the realization that proteins could not be made directly from DNA, since in eukaryotes DNA resides within the nucleus, whereas protein synthesis had been observed to happen outside the nucleus, in the *cytoplasm*. Therefore, some unknown agent had to somehow transport the genetic information from the DNA in the nucleus to the cytoplasm. In the mid 1950s Paul Zamecnik discovered that protein synthesis in the cytoplasm happens with the help of certain large molecules called *ribosomes* that contain RNA. This led to the suspicion that RNA could be the intermediary agent between DNA and proteins. Finally, in 1960 Benjamin Hall and Sol Spiegelman demonstrated that RNA forms duplexes with single-stranded DNA, proving that the RNA (responsible for the synthesis of a particular protein) is complementary to the DNA segment (i.e., the gene) that codes for the protein. Thus, DNA served as a template used to copy a particular gene into *messenger RNA* (*mRNA*) that carries the gene's genetic information to the ribosome to make a particular protein.

Chemically speaking, RNA, or *ribonucleic acid*, is almost the same as DNA. There are two main differences between RNA and DNA: there is no T base in RNA—the similar base U takes its place—and an oxygen atom is added to the sugar component. These two seemingly minor differences have a major impact on the biological roles of the two molecules. DNA is mostly inert and almost always double-stranded, helping it to serve as a static repository for information. RNA, on the other hand, is more chemically active and it usually lives in a single-stranded form. The effect is that RNA can carry short messages from the DNA to the cellular machinery that builds protein, and it can actively participate in important chemical reactions.

In 1960 Jerard Hurwitz and Samuel Weiss identified a molecular machine (composed of many proteins) that uses DNA as a template and adds ribonucleotide by ribonucleotide to make RNA. This process is called *transcription* and the molecular machine responsible for this process got the name *RNA polymerase*. Despite the advances in our understanding of the copying of DNA into RNA, how RNA polymerase knows where to start and stop transcribing DNA remains one of the many unsolved bioinformatics problems. Furthermore, the transcription of a gene into mRNA is tightly controlled, so that not all genes produce proteins at all times. Though some basic mechanisms of how gene transcription is controlled are known, a comprehensive understanding for all genes is still beyond our grasp.

In eukaryotes, a gene is typically broken into many pieces but it still produces a coherent protein. To do so, these cells have to cut the introns out of the RNA transcript and concatenate all the exons together prior to the mRNA entering the ribosome. This process of cutting and pasting the "raw" RNA version of the gene into the mRNA version that enters the ribosome is called *splicing* and is quite complicated at a molecular level.

How are Proteins Made?

In 1820 Henry Braconnot identified the first amino acid, glycine. By the early 1900s all twenty amino acids had been discovered and their chemical structure identified. Since the early 1900s when Emil Hermann Fischer showed that amino acids were linked together into linear chains to form proteins, proteins became the focus of biochemistry and molecular biology. It was postulated that the properties of proteins were defined by the composition and arrangement of their amino acids, which we now accept as true.

To uncover the code responsible for the transformation of DNA into protein, biologists conjectured that triplets of consecutive letters in DNA (called *codons*) were responsible for the amino acid sequence in a protein. Thus, a particular 30-base pair gene in DNA will make a protein of a specific 10 amino acids in a specific order. There are $4^3 = 64$ different codons, which is more than three times as large as the number of amino acids. To explain this redundancy biologists conjectured that the *genetic code* responsible for transforming DNA into protein is degenerate: different triplets of nucleotides may code for the same amino acid. Biologists raced to find out which triplets code for which amino acids and by the late 1960s discovered the *genetic code*. The *triplet rule* was therefore confirmed and is now accepted as fact.

Unlike the regular double-helical structure of DNA, the three-dimensional structure of proteins is highly variable. Researchers invest a large amount of effort into finding the structure of each protein; it is this structure that determines what role a protein plays in the cell—does it participate in the DNA replication process, or does it take part in some pathway that helps the cell metabolize sugar faster? Proteins perform most of the chemical work in the cell, including copying DNA, moving materials inside the cell, and communicating with nearby cells. Biologists used to believe that one gene coded for one protein, but a more complex picture emerged recently with the

DNA:	TAC	CGC	GGC	TAT	TAC	TGC	CAG	GAA	GGA	ACT
RNA:	AUG	GCG	CCG	AUA	AUG	ACG	GUC	CUU	CUU	UAG
Protein:	Met	Ala	Pro	Ile	Met	Thr	Val	Leu	Pro	Stop

Fig. 3.1. The transcription of DNA into RNA, and the translation of RNA into a protein. Every amino acid is denoted with three letters, for example Met stands for the amino acid Methionine.

Table 3.1. The genetic code from the perspective of mRNA.

	U		C		A		G	
U	UUU	Phe	UCU	Ser	UAU	Tyr	UGU	Cys
	UUC	Phe	UCC	Ser	UAC	Tyr	UGC	Cys
	UUA	Leu	UCA	Ser	UAA	Stop	UGA	Stop
	UUG	Leu	UCG	Ser	UAG	Stop	UGG	Trp
C	CUU	Leu	CCU	Pro	CAU	His	CGU	Arg
	CUC	Leu	CCC	Pro	CAC	His	CGC	Arg
	CUA	Leu	CCA	Pro	CAA	Gln	CGA	Arg
	CUG	Leu	CCG	Pro	CAG	Gln	CGG	Arg
A	AUU	Ile	ACU	Thr	AAU	Asn	AGU	Ser
	AUC	Ile	ACC	Thr	AAC	Asn	AGC	Ser
	AUA	Ile	ACA	Thr	AAA	Lys	AGA	Arg
	AUG	Met	ACG	Thr	AAG	Lys	AGG	Arg
G	GUU	Val	GCU	Ala	GAU	Asp	GGU	Gly
	GUC	Val	GCC	Ala	GAC	Asp	GGC	Gly
	GUA	Val	GCA	Ala	GAA	Glu	GGA	Gly
	GUG	Val	GCG	Ala	GAG	Glu	GGG	Gly

discovery of *alternative splicing*, allowing one gene to code for many proteins.

Many chemical systems in the cell require *protein complexes*, which are groups of proteins that clump together into a large structure. A protein complex, known as RNA polymerase, begins *transcribing* a gene by copying its DNA base sequence into a short RNA base sequence (pairing a DNA T with an RNA A, a DNA A with an RNA U, and so on) called *messenger RNA*, or mRNA. This short molecule is then attacked by large molecular complexes known as *ribosomes*, which read consecutive codons and locate the corresponding amino acid for inclusion in the growing polypeptide chain. Ribosomes are, in effect, molecular factories where proteins are assembled.

To help with the location of the proper amino acid for a given codon, a special type of RNA, called *transfer RNA* (tRNA), performs a specific and elegant function. There are twenty types of tRNAs, and twenty types of amino acids. Each type of amino acid binds to a different tRNA, and the tRNA molecules have a three-base segment (called an *anticodon*) that is complementary to the codon in the mRNA. As in DNA base-pairing, the anticodon on the tRNA sticks to the codon on the RNA, which makes the amino acid available to the ribosome to add to the polypeptide chain. When one amino acid has been added, the ribosome shifts one codon to the right, and the process repeats. The process of turning an mRNA into a protein is called *translation*, since it translates information from the RNA (written in a four-letter alphabet) into the protein (written in 20-letter alphabet). All proteins, including the ones necessary for this process, are produced by this process.

This flow of information,

DNA → *transcription* → RNA → *translation* → protein,

is emphatically referred to as the *central dogma in molecular biology*.

How can We Analyze DNA?

Over the years, biologists have learned how to analyze DNA. Below we describe some important techniques for copying, cutting, pasting, measuring, and probing DNA.

Copying DNA

Why does one need to copy DNA, that is, to obtain a large number of identical DNA fragments? From a computer science perspective, having the same string in 10^9 copies does not mean much since it does not increase the total amount of information. However, most experimental techniques (like gel electrophoresis, used for measuring DNA length) require many copies of the same DNA fragment. Since it is difficult to detect a single molecule or even a hundred molecules with modern instrumentation, amplifying DNA to yield millions or billions of identical copies is often a prerequisite of further analysis.

PCR amplifies a short (100- to 500-nucleotide) DNA fragment and produces a large number of identical DNA strings. To use PCR, one must know a pair of short (20- to 30-letter) strings in the DNA flanking the area of interest and design two *PCR primers*, synthetic DNA fragments identical to these strings.

Suppose we want to generate a billion copies of a DNA fragment of 500 nucleotides, that we know happens to be flanked by the 20-mer nucleotide sequence X on the left and the 20-mer nucleotide sequence Y on the right. PCR repeats a cycle of three operations: *denaturation*, *priming*, and *extension* to double the number of DNA fragments in every iteration. Therefore, after thirty iterations of PCR we will have on the order of 2^{30} DNA fragments, which is more than a billion copies. To start PCR, we only need a single copy of the target DNA, some artificially synthesized 20-nucleotide long DNA fragment $\overline{X}$ (many copies), some 20-nucleotide long DNA fragment Y (many copies), and billions of "spare" nucleotides (A,T,G,C). We also need a molecular machine that will copy an existing DNA strand to produce a new DNA strand, and for this purpose we hijack DNA polymerase. DNA polymerase has an ability to add a complementary copy to a single-stranded DNA as long as there is a primer (i.e., $\overline{X}$ and Y) attached to the DNA strand and a sufficient supply of spare nucleotides

The denaturation step simply amounts to heating double-stranded DNA to separate it into two single strands. Priming is cooling down

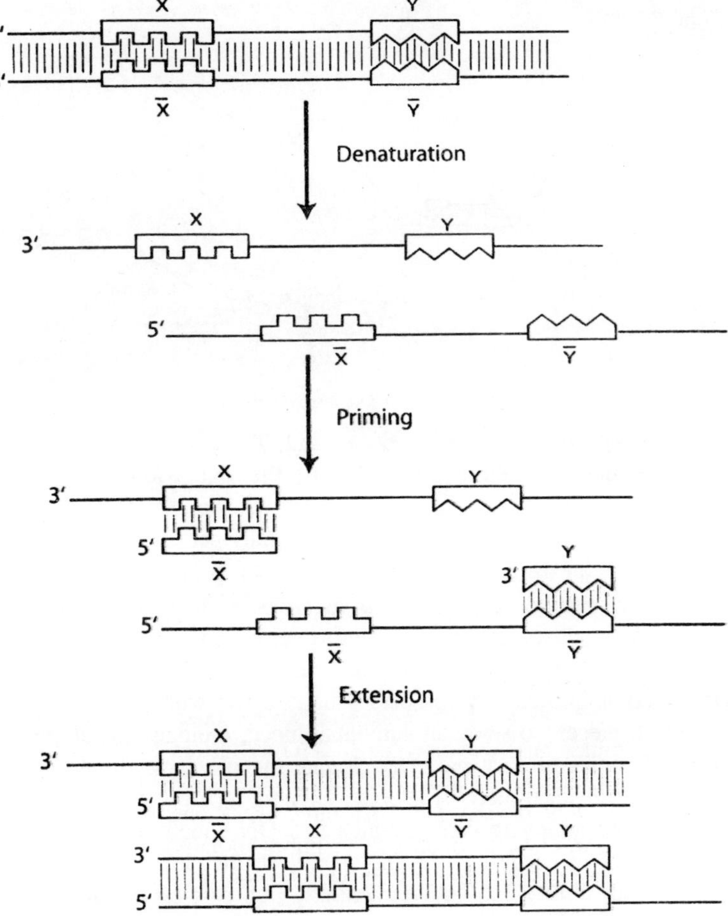

Fig. 3.2. The three main operations in the polymerase chain reaction.

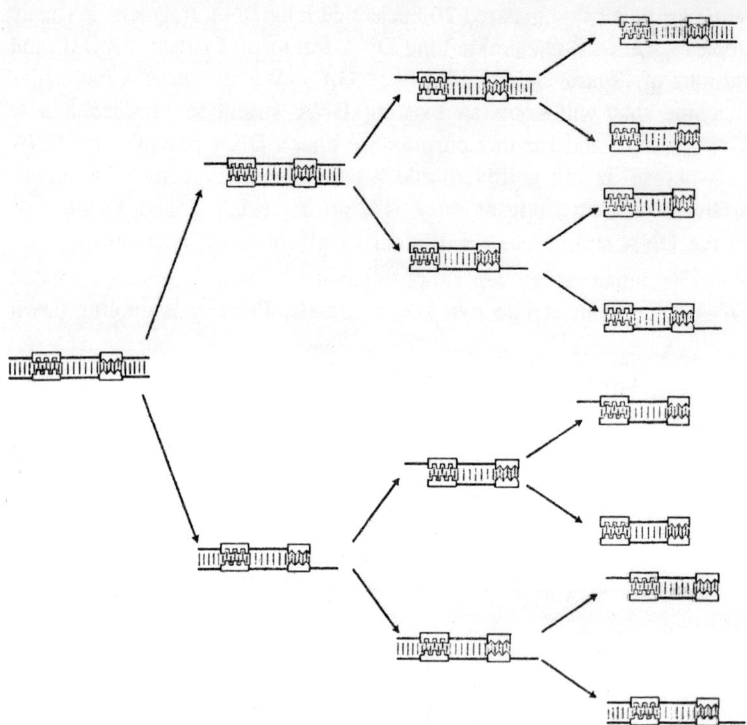

Fig. 3.3. The first few iterations of PCR.

the solution to allow primers $\overline{X}$ and Y to hybridize to their complementary positions in DNA. In the extension step, DNA polymerase extends the primer to produce two double-stranded DNA copies from single-stranded DNA. By repeatedly performing these three steps, one achieves an exponential increase in the amount of DNA.

Another way to copy DNA is to *clone* it. In contrast to PCR, cloning does not require any prior information about flanking primers. However, biologists usually have no control over *which* fragment of DNA gets amplified. The process usually starts with breaking DNA into small pieces; to study an individual piece, biologists obtain many identical copies of each piece by cloning the pieces, and then try to select the individual piece of interest. Cloning incorporates a fragment of DNA into a *cloning vector*, which is a DNA molecule originating from a virus or bacterium. In this operation, the cloning vector does not lose its ability for self-replication, but carries the additional incorporated *insert* that the biologist plans to study. Vectors introduce foreign DNA into host cells (such as bacteria) which reproduce in large quantities. The self-replication process creates a large number of copies of the fragment, thus enabling its properties to be studied. A fragment reproduced in this way is called a *clone*. Biologists can make *clone libraries* consisting of thousands of clones (each representing a short, randomly chosen DNA fragment) from the same DNA molecule. For example, the entire human genome can be represented as a library of 30,000 clones, each clone carrying a 100- to 200-kilobase (1000 base pairs) insert.

Cutting and Pasting DNA

In order to study a gene (more generally, a genomic region) of interest, it is sometimes necessary to cut it out of an organism's genome and reintroduce it into some host organism that is easy to grow, like a bacterium. Fortunately, there exist "scissors" that do just this task: certain proteins destroy the internal bonds in DNA molecules, effectively cutting it into pieces. *Restriction enzymes* are proteins that act as molecular scissors that cut DNA at every occurrence of a certain string (recognition site). For example, the *Bam*HI restriction enzyme cuts DNA into *restriction fragments* at every occurrence of the string GGATCC. Restriction enzymes first bind to the recognition site in the double-stranded DNA and then cut the DNA. The cut may produce blunt or sticky ends. Biologists have many ways to fuse two pieces of DNA together by adding the required chemical bonds. This is usually done by mimicking the processes that happen in the cell all the time.

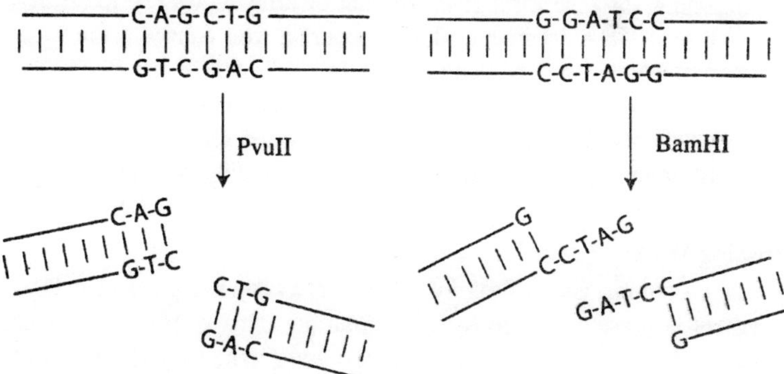

Fig. 3.4. Sticky and blunt ends after cutting DNA with restriction enzymes. BamHI and PvuII cut at GGATCC and CAGCTG, respectively, both of which are palindromes. However, the result of BamHI leaves four unmatched nucleotides on each of the strands that are cut; if a gene is cut out of one organism with BamHI, it can be inserted into a different sequence that has also been cut with BamHI because the sticky ends act as glue.

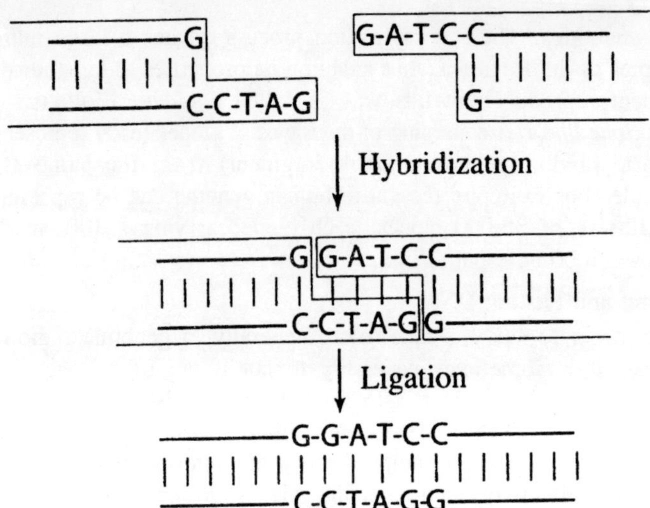

Fig. 3.5. Cutting and pasting two fragments that have sticky ends.

Measuring DNA Length

Gel electrophoresis is a technique that allows a biologist to measure the size of a DNA fragment without actually finding its exact sequence. DNA is a negatively charged molecule that migrates toward the positive pole of an electric field. The gel acts as a molecular "brake" so that long molecules move slower than short ones. "The speed of migration of a fragment is related to the fragment's size, so the measurement of the migration distance for a given amount of time allows one to estimate the size of a DNA fragment. But, of course, you cannot actually see DNA molecules, so "molecular light bulbs," which are fluorescent compounds, are attached by a chemical reaction to the ends of the DNA fragments. With these bulbs, biologists can see how far different DNA fragments in a mixture migrate in the gel and thus estimate their respective lengths.

Probing DNA

A common task in biology is to test whether a particular DNA fragment is present in a given DNA solution. This is often done using *hybridization*: the process of joining two complementary DNA strands into a single double-stranded molecule. Biologists often use *probes*, which are single-stranded DNA fragments 20 to 30 nucleotides long that have a known sequence and a fluorescent tag. Hybridization of the probe to some unknown DNA fragment of interest can show a biologist the presence of the probe's complementary sequence in the larger DNA fragment. We can also probe RNA using a *DNA array* to see if a gene is on or off. A DNA array is essentially composed of "spots" bound to a solid support, such as a glass slide. On each spot are many copies of the complement of one gene's mRNA transcript. If the mRNA content of a cell is poured onto this slide, the mRNA will bind to the single-stranded spots and can be detected with the light-bulb technique described earlier. As a result, biologists can find out which genes are producing mRNA in a particular tissue under fixed conditions.

How do Individuals of a Species Differ?

The genetic makeup of an individual manifests itself in *traits*, such as hair colour, eye colour, or susceptibility to malaria. Traits are caused by variations in genes. A surprising observation is that, despite the near similarity of genomes among all humans, no two individuals are quite the same. In fact, the variations among the same gene across different individuals are limited to a handful of different base pairs (if any). Roughly only 0.1% of the 3 billion nucleotide human genome (or 3 million bases) are different between any two individuals. Still, this leaves room for roughly $4^{3,000,000}$ different genomes, and is for all intents and purposes an endless diversity.

In other words, when we speak of "the" genome of a species, we are referring to some sort of "master" genome that is fairly representative of all the possible genomes that an individual of that species could have. While specific individuals of the species may differ in some bases, the basic long DNA sequence is roughly the same in all members of the species. Of course, this handful of differences is critically important, and the large Human Diversity Project is underway to understand how various individuals differ. This will hopefully identify the mutations responsible for a number of genetic diseases.

How do Different Species Differ?

The genomes of different organisms may be vastly different and amazingly similar. The human genome consists of about 3 billion bases, while the fly genome has a scant 140 million bases. However, an analysis of the genomic sequences for two vastly different organisms (fruit flies and humans) has revealed that many genes in humans and flies are similar. Moreover, as many as 99% of all human genes are conserved across all mammals! Some human genes show strong similarity across not only mammals and flies but also across worms, plants, and (worse yet) deadly bacteria. A species, then, is a collection of individuals whose genomes are "compatible," in the sense of mating.

The likelihood that all currently living species could spontaneously develop the same gene with the same function is quite low, so it seems reasonable to assume that some process must exist that generates new species from old ones. This process is called *evolution*. The theory that all living things have evolved through a process of incremental change over millions of years has been at the heart of biology since the publication in 1859 of Charles Darwin's *On the Origin of Species*. However, only with the discovery of genomic sequences were biologists able to see how these changes are reflected in the genetic texts of existing species.

There are a number of sources for genetic variation across individuals in a species. Errors in the replication of DNA, and bizarre biological processes such as reverse transcription all cause the genomes of any two individuals in a species to be subtly different. However, genetic differences are not entirely spurious; many organisms have inherent processes that enforce genetic variation, so that no two individuals could be the same. Occasionally, a variation in an individual's genome can produce a new trait, perhaps slightly stronger teeth or longer fins. If the mutations in an individual are beneficial in that individual's environment, then that individual will be more likely to be reproductively successful, passing along the mutation to its progeny. If the mutations are harmful, then that individual will be less likely to reproduce and the mutation will die out. This filtering of mutations is called *natural selection*. Over many generations the more successful individuals will become an increasingly large part of the population, to the end that the beneficial mutation gradually takes root in all the living members of a species. As the species, as a whole, takes on the new trait, we say that it *adapts* to its environment.

If a species is divided into two isolated groups and placed into different environments, then the groups will adapt differently. After many more generations, the two groups become so different that their individuals can no longer reproduce with each other, and they have become different species. This process is called *speciation*. Adaptation and speciation together form the basis of the process of evolution by natural selection, and explains the apparent paradox that there is such a diversity of life on the planet, yet so many of the genes seem similar at a sequence level. The recent abundance of genomic sequence data has enabled bioinformaticians to carry out studies that try to unravel the evolutionary relationships among different species. Since evolution by natural selection is the direct effect of adjustments to a species' genomic sequence, it stands to reason that studying the genomic sequences in different species can yield insight into their evolutionary history.

WHY BIOINFORMATICS?

As James Watson and Francis Crick worked to decipher the DNA puzzle, the 30 year-old English architect Michael Ventris tried to decipher an ancient language known as *Linear B*. At the beginning of the twentieth century, archaeologists excavated the ancient city of Knossos located on the island of Crete and found what might have been the palace of King Minos, complete with labyrinth. The archaeologists also found clay tablets with an unfamiliar form of writing. These were letters of an unknown language and there was nothing to compare them to.

The script that the ancient Cretans used (nicknamed "Linear B") remained a mystery for the next fifty years. Linguists at that time thought that Linear B was used to write in some hypothetical Minoan language (i.e., after King Minos) and cut off any investigation into the possibility that the language on the tablets was Greek. In 1936, a fourteen-year-old boy, Michael Ventris, went on a school trip to the Minoan exhibit in London and was fascinated with the legend of the Minotaur and the unsolved puzzle of the Minoan language. After seventeen years of code-breaking, Ventris decoded the Minoan language at about the same time Watson and Crick deciphered the structure of DNA.

Some Linear B tablets had been discovered on the Greek mainland. Noting that certain strings of symbols appeared in the Cretan texts but did not appear in Greek texts, Ventris made the inspired guess that those strings applied to cities on the island. Armed with these new symbols that he could decipher, he soon unlocked much more text, and determined that the underlying language of Linear B was, in fact, just Greek written in a different alphabet. This showed that the Cretan civilization of the Linear B tablets had been part of Greek civilization.

There were two types of clay tablets found at Crete: some written in Linear B and others written in a different script named *Linear A*. Linear A appears to be older than Linear B and linguists think that Linear A is the oldest written language of Europe, a precursor of Greek. Linear A has resisted all attempts at decoding. Its underlying language is still unknown and probably will remain undecoded since it does not seem to relate to any other surviving language in the world.

Linear A and Linear B texts are written in alphabets consisting of roughly ninety symbols.

Bioinformatics was born after biologists discovered how to sequence DNA and soon generated many texts in the four-letter alphabet of DNA. DNA is more like Linear A than Linear B when it comes to decoding—we still know very little about the language of DNA. Like Michael Ventris, who mobilized the mathematics of code-breaking to decipher Linear B, bioinformaticians use algorithms, statistics, and other mathematical techniques to decipher the language of DNA.

For example, suppose we have the genomic sequences of two insects that we suspect are somewhat related, evolutionarily speaking—perhaps a fruit fly (*Drosophila melanogaster*) and a malaria mosquito (*Anopheles gamibae*). Taking the Michael Ventris approach, we would like to know what parts of the fruit fly genomic sequence are dissimilar and what parts are similar to the mosquito genomic sequence. Though the means to find this out may not be immediately obvious at this point, the alignment algorithms described later in this book allow one to compare any two genes and to detect similarities between them. Unfortunately, it will take an unbearably long time to do so if we want to compare the entire fruit fly genome with the entire mosquito genome. Rather than giving up on the question altogether, biologists combined their efforts with algorithmists and mathematicians to come up with an algorithm (BLAST) that solves the problem very quickly and evaluates the statistical significance of any similarities that it finds.

Comparing related DNA sequences is often a key to understanding each of them, which is why recent efforts to sequence many related genomes (e.g., human, chimpanzee, mouse, rat) provide the best hope for understanding the language of DNA. This approach is often referred to as *comparative genomics*. A similar approach was used by the nineteenth century French linguist Jean-Francois Champollion who decoded the ancient Egyptian language.

The ancient Egyptians used hieroglyphs, but when the Egyptian religion was banned in the fourth century as a pagan cult, knowledge of hieroglyphics was lost. Even worse, the spoken language of Egyptian and its script (known as *demotic*) was lost soon afterward and completely forgotten by the tenth century when Arabic became the language of Egypt. As a result, a script that had been in use since the beginning of the third millennium BC turned into a forgotten language that nobody remembered.

During Napoleon's Egyptian campaign, French soldiers near the city of Rosetta found a stone (now known as the *Rosetta stone*) that was inscribed in three different scripts. Many of Napoleon's officers happened to be classically educated and one of them, a Lieutenant Bouchard, identified the three bands of scripts as hieroglyphic, demotic, and ancient Creek. The last sentence of the Greek inscription read: "This decree shall be inscribed on stelae of hard rock, in sacred characters, both native and Greek." The Rosetta stone thus presented a comparative linguistics problem not unlike the comparative genomics problems bioinformaticians face today.

In recent decades biology has raised fascinating mathematical problems and has enabled important biological discoveries. Biologists that reduce bioinformatics to simply "the application of computers in biology" sometimes fail to recognize the rich intellectual content of bioinformatics. Bioinformatics has become a part of modern biology and often dictates new fashions, enables new approaches, and drives further biological developments. Simply using bioinformatics as a tool kit without a reasonable understanding of the main computational ideas is not very different from using a PCR kit without knowing how PCR works.

Bioinformatics is a large branch of biology (or of computer science) and this book presents neither a complete cross section nor a detailed look at any one part of it. Our intent is to describe those algorithmic principles that underlie the solution to several important biological problems to make it possible to understand any other part of the field.

4

Alignment of Pairs of Sequences

Definition of Sequence Alignment

Sequence alignment is the procedure of comparing two (pair-wise alignment) or more (multiple sequence alignment) sequences by searching for a series of individual characters or character patterns that are in the same order in the sequences. Two sequences are aligned by writing them across a page in two rows. Identical or similar characters are placed in the same column, and nonidentical characters can either be placed in the same column as a mismatch or opposite a gap in the other sequence. In an optimal alignment, nonidentical characters and gaps are placed to bring as many identical or similar characters as possible into vertical register. Sequences that can be readily aligned in this manner are said to be similar.

There are two types of sequence alignment, global and local. In global alignment, an attempt is made to align the entire sequence, using as many characters as possible, up to both ends of each sequence. Sequences that are quite similar and approximately the same length are suitable candidates for global alignment. In local alignment, stretches of sequence with the highest density of matches are aligned, thus generating one or more islands of matches or subalignments in the aligned sequences. Local alignments are more suitable for aligning sequences that are similar along some of their lengths but dissimilar in others, sequences that differ in length, or sequences that share a conserved region or domain.

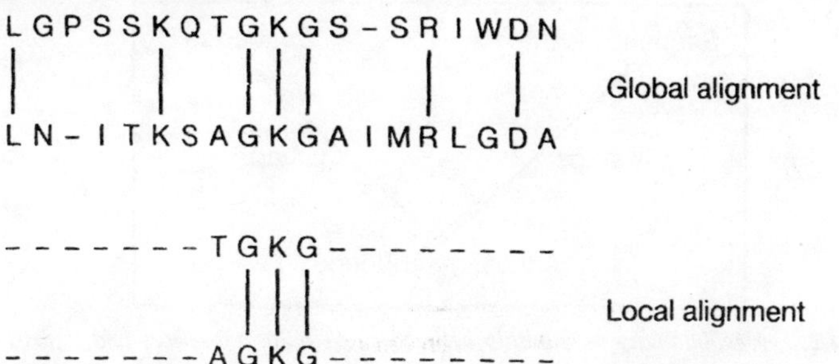

Fig. 4.1. Distinction between global and local alignments of two sequences.

Global Alignment

For the two hypothetical protein sequence, the global alignment is stretched over the entire sequence length to include as many matching amino acids as possible up to and including the sequence ends. Vertical bars between the sequences indicate the presence of identical amino acids. Although there is an obvious region of identity in this example (the sequence GKG preceded by a commonly observed substitution of T for A), a global alignment may not align such regions so that more amino acids along the entire sequence lengths can be matched.

Local Alignment

In a local alignment, the alignment stops at the ends of regions of identity or strong similarity, and a much higher priority is given to finding these local regions than to extending the alignment to include more neighbouring amino acid pairs. Dashes indicate sequence not included in the alignment. This type of alignment favours finding conserved nucleotide patterns, DNA sequences, or amino acid patterns in protein sequences.

Significance of Sequence Alignment

Sequence alignment is useful for discovering functional, structural, and evolutionary information in biological sequences. It is important to obtain the best possible or so-called "optimal" alignment to discover this information. Sequences that are very much alike, or "similar" in the parlance of sequence analysis, probably have the same function, be it a regulatory role in the case of similar DNA molecules, or a similar biochemical function and three-dimensional structure in the case of proteins. Additionally, if two sequences from different organisms are similar, there may have been a common ancestor sequence, and the sequences are then defined as being homologous. The alignment indicates the changes that could have occurred between the two homologous sequences and a common ancestor sequence during evolution.

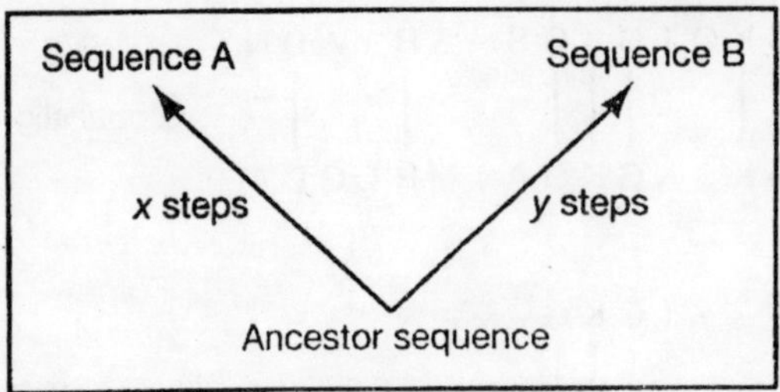

Fig. 4.2. The evolutionary relationship between two similar sequences and a possible common ancestor sequence that would make the sequences homologous.

With the advent of genome analysis and large-scale sequence comparisons, it becomes important to recognize that sequence similarity may be an indicator of several possible types of ancestor relationships, or there may be no ancestor relationship at all. For example, new gene evolution is often thought to occur by gene duplication, creating two tandem copies of the gene, followed by mutations in these copies. In rare cases, new mutations in one of the copies provide an advantageous change in function. The two copies may then evolve along separate pathways. Although the resulting separation of function will generate two related sequence families, sequences among both families will still be similar due to the single gene ancestor. In addition, genetic rearrangements can reassort domains in proteins, leading to more complex proteins with an evolutionary history that is difficult to reconstruct.

Evolutionary theory provides terms that may be used to describe sequence relationships. Homologous genes that share a common ancestry and function in the absence of any evidence of gene duplication are called *orthologs*. When there is evidence for gene duplication, the genes in an evolutionary lineage derived from one of the copies and with the same function are also referred to as orthologs. The two copies of the duplicated gene and their progeny in the evolutionary lineage are referred to as *paralogs*. In other cases, similar regions in sequences may not have a common ancestor but may have arisen independently by two evolutionary pathways converging on the same function, called *convergent evolution*. There are some remarkable examples in protein structures. For instance, although the enzymes chymotrypsin and subtilisin have totally different three-dimensional structures and folds, the active sites show similar structural features, including histidine (H), serine (S), and aspartic acid (D) in the catalytic sites of the enzymes. In such cases, the similarity will be highly localized. Such sequences are referred to as *analogous*. A closer examination of alignments can help to sort out possible evolutionary origins among similar sequences.

As pointed out by Fitch and Smith (1983), sequences can be either *homologous* or *non-homologous*, but not in between. The genetic rearrangements referred to above can give rise to chimeric genes, in which some regions are homologous and others are not. Referring to the entire sequences as homologous in such situations leads to an inaccurate and incomplete description of the sequence lineage.

Another complication in tracing the origins of similar sequences is that individual genes may not share the same evolutionary origin as the rest of the genome in which they presently reside. Genetic events such as symbioses and viral-induced transduction can cause horizontal transfer of genetic material between unrelated organisms. In such cases, the evolutionary history of the transferred sequences and that of the organisms will be different. Again, with the capability of detecting such events in the genomes of organisms comes the responsibility to describe these changes with the correct evolutionary terminology. In this case, the sequences are *xenologous*. Recently, Lawrence and Ochman (1997) have shown that horizontal transfer of genes between species is as common in enteric bacteria, if not more common, than mutation. Describing such changes requires a careful description of sequence origins.

Overview of Methods of Sequence Alignment

Alignment of Pairs of Sequences

Alignment of two sequences is performed using the following methods:

1. Dot matrix analysis
2. The dynamic programming (or DP) algorithm
3. Word or *k*-tuple methods, such as used by the programs FASTA and BLAST.

Unless the sequences are known to be very much alike, the dot matrix method should be used first, because this method displays any possible sequence alignments as diagonals on the matrix. Dot matrix analysis can readily reveal the presence of insertions/deletions and direct and inverted repeats that are more difficult to find by the other, more automated methods. The major limitation of the method is that most dot matrix computer programs do not show an actual alignment.

The dynamic programming method, first used for global alignment of sequences by Needleman and Wunsch (1970) and for local alignment by Smith and Waterman (1981 a), provides one or more alignments of the sequences. An alignment is generated by starting at the ends of the two sequences and attempting to match all possible pairs of characters between the sequences and by following a scoring scheme for matches, mismatches, and gaps. This procedure generates a matrix of numbers that represents all possible alignments between the sequences. The highest set of sequential scores in the matrix defines an optimal alignment. For proteins, an amino acid substitution matrix, such as the Dayhoff percent accepted mutation matrix 250 (PAM250) or blosum substitution matrix 62 (BLOSUM62) is used to score matches and mismatches. Similar matrices are available for aligning DNA sequences.

The dynamic programming method is guaranteed in a mathematical sense to provide the optimal (very best or highest-scoring) alignment for a given set of user-defined variables, including choice of scoring matrix and gap penalties. Fortunately, experience with the dynamic programming method has provided much help for making the best choices, and dynamic programming has become widely used. The dynamic programming method can also be slow due to the very large number of computational steps, which increase approximately as the square or cube of the sequence lengths. The computer memory requirement also increases as the square of the sequence lengths. Thus, it is difficult to use the method for very long sequences. Fortunately, computer scientists have greatly reduced these time and space requirements to near-linear relationships without compromising the reliability of the dynamic programming method, and these methods are widely used in the available dynamic programming applications to sequence alignment. Other shortcuts have been developed to speed up the early phases of finding an alignment.

The word or *k*-tuple methods are used by the FASTA and BLAST algorithms. They align two sequences very quickly, by first searching for identical short stretches of sequences (called words or *k*-tuples) and by then joining these words into an alignment by the dynamic programming method. These methods are fast enough to be suitable for searching an entire database for the sequences that align best with an input test sequence. The FASTA and BLAST methods are heuristic; i.e., an empirical method of computer programming in which rules of thumb are used to find solutions and feedback is used to improve performance. However, these methods are reliable in a statistical sense, and usually provide a reliable alignment.

Multiple Sequence Alignment

From a multiple alignment of three or more protein sequences, the highly conserve residues that define structural and functional domains in protein families can be identified. New members of such families can then be found by searching sequence databases for other sequences with these same domains. Alignment of DNA sequences can assist in finding conserved regulatory patterns in DNA sequences. Despite the great value of multiple sequence alignments, obtaining one presents a very difficult algorithmic problem.

DOT MATRIX SEQUENCE COMPARISON

A dot matrix analysis is primarily a method for comparing two sequences to look for possible alignment of characters between the sequences, first described by Gibbs and McIntyre (1970). The method is also used for finding direct or inverted repeats in protein and DNA sequences, and for predicting regions in RNA that are self-complementary and that, therefore, have the potential of forming secondary structure. Every laboratory that does sequence analysis should have at least one dot matrix program available. In choosing a program, look for as many of the features described below as possible. The dot matrix should be visible on the computer terminal, thus providing an interactive environment so that different types of analyses may be tried. Use of coloured dots can enhance the detection of regions of similarity. Additional descriptions of the dot matrix method have appeared elsewhere. The examples given below use the dot matrix module of DNA Strider (version 1.3) on a Macintosh computer. The program DOTTER has interactive features for the UNIX X-Windows environment. The Genetics Computer Group programs COMPARE and DOTPLOT also perform a dot matrix analysis. Although not a dot matrix method, the program PLALIGN in the FASTA suite may be used to display the alignments found by the dynamic programming method between two sequences on a graph.

Pair-wise Sequence Comparison

The major advantage of the dot matrix method for finding sequence alignments is that all possible matches of residues between two sequences are found, leaving the investigator the choice of identifying the most significant ones. Then, sequences of the actual regions that align can be detected by using one of two other methods for performing sequence alignments, e.g., dynamic programming. These methods are automatic and usually show one best or optimal alignment, even though there may be several different, nearly alike alignments. Alignments

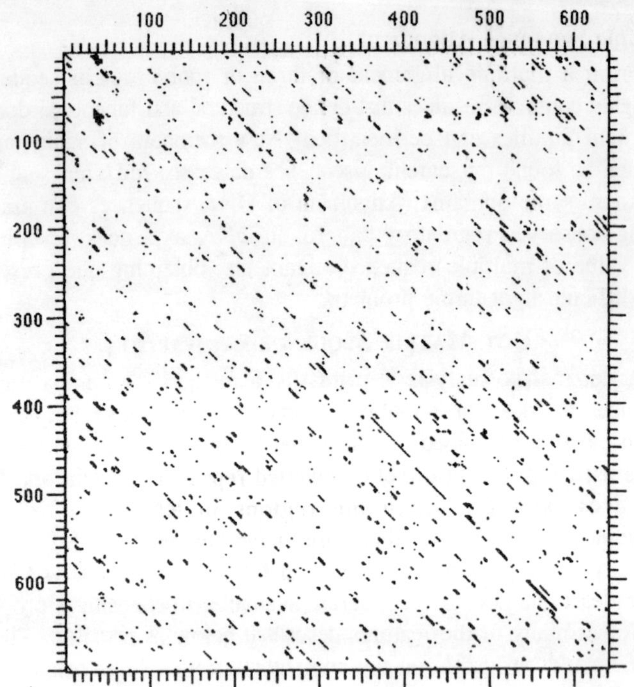

Fig. 4.3. Dot matrix analysis of DNA sequences encoding phage λ cI and phage P22 c2.

generated by these programs can he compared to the dot matrix alignment to determine whether the longest regions are being matched and whether insertions and deletions are located in the most reasonable places.

In the dot matrix method of sequence comparison, one sequence (A) is listed across the top of a page and the other sequence (B) is listed down the left side. Starting with the first character in B, one then moves across the page keeping in the first row and placing a dot in any column where the character in A is the same. The second character in B is then compared to the entire A sequence, and a dot is placed in row 2 wherever a match occurs. This process is continued until the page is filled with dots representing all the possible matches of A characters with B characters. Any region of similar sequence is revealed by a diagonal row of dots. Isolated dots not on the diagonal represent random matches that are probably not related to any significant alignment.

Detection of matching regions may be improved by filtering out random matches in a dot matrix. Filtering is achieved by using a

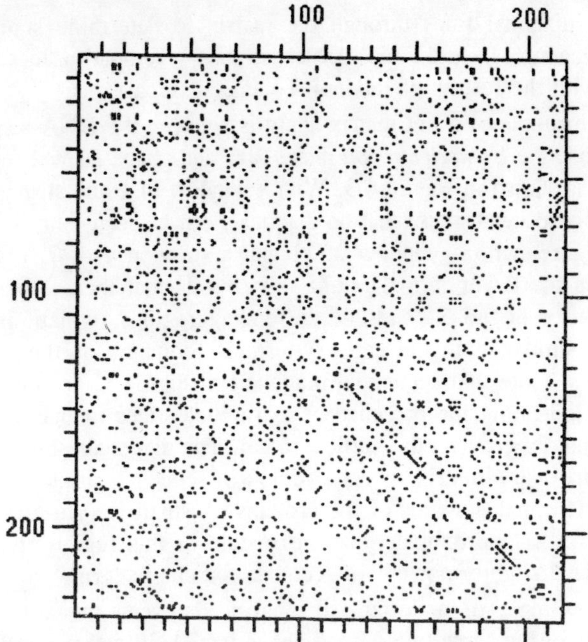

Fig. 4.4. Dot matrix analysis of the amino acid sequences of the phage λ cI and phage P22 c2.

sliding window to compare the two sequences. Instead of comparing single sequence positions, a window of adjacent positions in the two sequences is compared at the same time, and a dot is printed on the page only if a certain minimal number of matches occur. The window starts at the positions in A and B to be compared and includes characters in a diagonal line going down and to the right, comparing each pair in turn, as in making an alignment. A larger window size is generally used for DNA sequences than for protein sequences because the number of random matches is much greater due to the use of only four DNA symbols as compared to 20 amino acid symbols. A typical window size for DNA sequences is 15 and a suitable match requirement in this window is 10. For protein sequences, the matrix is often not filtered, but a window size of 2 or 3 and a match requirement of 2 will highlight matching regions. If two proteins are expected to be related but to have long regions of dissimilar sequence with only a small proportion of identities, such as similar active sites, a large window, e.g., 20, and small stringency, e.g., 5, should be useful for seeing any similarity. Identification of sequence alignments by the dot matrix method can be aided by performing a count of dots in all possible diagonal lines through the matrix to determine statistically which diagonals have the most matches, and by comparing these match scores with the results of random sequence comparisons.

An example of a dot matrix analysis between the DNA sequences that encode the *Escherichia coli* phage λ *cI* and phage P22 *c2* repressor proteins is shown in Figure 4.3. With a window of 1 and stringency of 1, there is so much noise that no diagonals can be seen, but, as shown in the figure, with a window of 11 and a stringency of 7, diagonals appear in the lower left. The analysis reveals that there are regions of similarity in the 3′ ends of the coding regions, which, in turn, suggests similarity in the carboxy-terminal domains of the encoded repressors. Note that sequential diagonals in matrix C do not line up exactly, indicating the presence of extra nucleotides in one sequence (the lambda *cI* gene on the vertical scale). The diagonals shown in the lower part of the matrix reveal a region of sequence similarity in the carboxy-terminal domains of the proteins. A small insertion in the cI protein that is approximately in the middle of this region and shifts the diagonal slightly downward accounts for this pattern.

An example of a dot matrix analysis between the amino acid sequences of the same two *E. coli* phage lambda cI and phage P22 *c2* repressor proteins is shown in Figure 4.4. This matrix was filtered by a window of 1 and a stringency of 1. As found with the DNA sequence alignment of the corresponding genes, diagonals shown in the lower part of the matrix reveal a region of sequence similarity in the carboxy-terminal domains of the proteins. The small insertion in the cI protein approximately in the middle of this region which shifts the diagonal slightly downward and which is also observed in the DNA alignment of these corresponding genes is also visible. Note that these windows are much smaller than required for DNA sequence comparisons due to the greater number of possible symbols (20 amino acids) and therefore fewer random matches.

In conclusion, for DNA sequence dot matrix comparisons, use long windows and high stringencies, e.g., 7 and 11, 11 and 15. For protein sequences, use short windows, e.g., 1 and 1, for window and stringency, respectively, except when looking for a short domain of partial similarity in otherwise not-similar sequences. In this case, use a longer window and a small stringency, e.g., 15 and 5, for window and stringency, respectively.

There are three types of variations in the analysis of two protein sequences by the do matrix method. First, chemical similarity of the amino acid R group or some other feature for distinguishing amino acids may be used to score similarity. Second, a symbol comparison table such as the PAM250 or BLOSUM62 tables may be used. These tables provide scores for matches based on their occurrence in aligned protein families. When these tables are used, a dot is placed in the matrix only if a minimum similarity score is found. These table values may also be used in a sliding window option, which averages the score within the window and prints a dot only above a certain average score. Finally, several different matrices can be made, each with a different scoring system, and the scores can be averaged. This method should be useful for aligning more distantly relates proteins. The scores of each possible diagonal through the matrix are then calculated, and the most significant ones are identified and shown on a computer screen.

Sequence Repeat

Dot matrix analysis can also be used to find direct and inverted repeats within sequences. Repeated regions in whole chromosomes may be detected by a dot matrix analysis, and an interactive Web-based program has been designed for showing these regions at increasing levels of detail. Direct repeats may also be found by performing sequence alignments with dynamic programming methods. When used to align a sequence with itself, the program LALIGN will show alternative possible alignments between the repeated regions; PLALIGN will plot these alignments on a graph similar in appearance to a dot matrix. Here, the sequence is analyzed against itself and the presence of repeats is revealed by diagonal rows of dots. Inverted repeats require special handling. A list of additional proteins with direct repeats is given in Doolittle, and repeats are also discussed in States and Boguski.

Repeats of a Single Sequence Symbol

A dot matrix analysis can also reveal the presence of repeats of the same sequence character many times. These repeats become apparent on the dot matrix of a protein sequence against itself as horizontal or vertical rows of dots that sometimes merge into rectangular or square patterns. Such patterns are particularly apparent in the right and lower regions of the dot matrix of the human LDL receptor but are also seen throughout the rest of the matrix. The occurrence of such repeats of the same sequence character increases the difficulty of aligning sequences because they create alignments with artificially high scores. A similar problem occurs with regions in which only a few sequence characters are found, called *low-complexity* regions.

Dynamic Programming Algorithm for Sequence Alignment

Dynamic programming is a computational method that is used to align two protein or nucleic acid sequences. The method is very important for sequence analysis because it provides the very best or optimal alignment between sequences. Programs that perform this analysis on sequences are readily available, and there are Web sites that will perform the analysis. However, the method requires the intelligent use of several variables in the program. Thus, it is important to understand how the program works in order to make informed choices of these variables.

The method compares every pair of characters in the two sequences and generates an alignment. This alignment will include matched and mismatched characters and gaps in the two sequences that are positioned so that the number of matches between identical or related characters is the maximum possible. The dynamic programming algorithm provides a reliable computational method for aligning DNA and protein sequences. The method has been proven mathematically to produce the best or optimal alignment between two sequences under a given set of match conditions. Optimal alignments provide useful information to biologists concerning sequence relationships by giving the best possible information as to which characters in a sequence should be in the same column in an alignment, and which are insertions in one of the sequences (or deletions on the other). This information is important for making functional, structural, and evolutionary predictions on the basis of sequence alignments.

Both global and local types of alignments may be made by simple changes in the basic dynamic programming algorithm. A global alignment program is based on the Needleman-Wunsch algorithm, and a local alignment program on the Smith-Waterman algorithm, described below. The predicted alignment will be given a score that gives the odds of obtaining the score between sequences known to be related to that obtained by chance alignment of unrelated sequences. There is a method to calculate whether or not an alignment obtained this way is statistically significant. One of the sequences may be scrambled many times and each randomly generated sequence may be realigned with the second sequence to demonstrate that the original alignment is unique.

Another feature of the dynamic programming algorithm is that the alignments obtained depend on the choice of a scoring system for comparing character pairs and penalty scores for gaps. For protein sequences, the simplest system of comparison is one based on identity. A match in an alignment is only scored if the two aligned amino acids are identical. However, one can also examine related protein sequences that can be aligned easily and find which amino acids are commonly substituted for each other. The probability of a substitution between any pair of the 20 amino acids may then be used to produce alignments. Recent improvements and experience with the dynamic programming programs and the scoring systems have greatly simplified their use.

It is important to recognize that several different alignments may provide approximately the same alignment score; i.e., there are alignments almost as good as the highest-scoring one reported by the alignment program. Some programs, e.g., LALIGN, provide several entirely different alignments with different sequence positions matched that can be compared to improve confidence in the best-scoring one. Alignment programs have also been greatly improved in algorithmic design and performance. With the advent of faster machines, it is possible to do a dynamic programming alignment between a query sequence and an entire sequence database and to find the similar sequences in several minutes. Dynamic programming has also been used to perform multiple sequence alignment, but only for a small number of sequences because the complexity of the calculations increases substantially for more than two sequences. Sequence alignment programs are available as a part of most sequence analysis packages, such as the widely used Genetics Computer Group GAP (global alignment) and BESTFIT (local alignment) programs. Sequences can also be pasted into a text area on a guest Web page on a remote host machine that will perform a dynamic programming alignment, and there are also versions of alignment programs that will run on a microcomputer.

In deciding to perform a sequence alignment, it is important to keep the goal of the analysis in mind. Is the investigator interested in trying to find out whether two proteins have similar domains or structural features, whether they are in the same family with a related biological function, or whether they share a common ancestor relationship? The desired objective will influence the way the analysis is done. There are several decisions to be made along the way, including the type of program, whether to produce a global or local alignment, the type of scoring matrix, and the value of the gap penalties to be used. There are a very large number of amino acid scoring matrices in use, some much more popular than others, and these scoring matrices are designed for different purposes. Some, such as the Dayhoff PAM matrices, are based on an evolutionary model of protein change, whereas others, such as the BLOSUM matrices, are designed to identify members of the same family. Alignments between DNA sequences require similar kinds of considerations. It is often worth the effort to try several approaches to find out which choice of scoring system and gap penalty give the most reasonable result. Fortunately, most alignment programs come with a recommended scoring matrix and gap penalties that are useful for most situations. A more recent development is the simultaneous use of a set of scoring matrices and gap penalties by a method that generates the most probable alignments. The final choice as to the most believable alignment is up to the investigator, subject to the condition that reasonable decisions have been made regarding the methods used.

For sequences that are very similar, e.g., >95%, the sequence alignment is usually quite obvious, and a computer program may not even be needed to produce the alignment. As the sequences become less and less similar, the alignment becomes more difficult to produce and one is less confident of the result. For protein sequences, similarity can still be recognized down to a level of approximately 25% amino acid identity. At this level of identity, the relative numbers of mismatched amino acids and gaps in the alignment have to be decided empirically and a decision made as to which gap penalties work the best for a given scoring matrix. Alignment of sequences at this level of identity is called the "twilight zone" of sequence alignment by Doolittle (1981). The alignment program may provide a quite convincing alignment, which suggests that the two sequences are homologous. The statistical significance of the alignment score may then be evaluated, as described later in this chapter.

Description of the Algorithm

Alignment of two sequences without allowing gaps requires an algorithm that performs a number of comparisons roughly proportional to the square of the average sequence length, as in a dot matrix comparison. If the alignment is to include gaps of any length at any position in either sequence, the number of comparisons that must be made becomes astronomical and is not achievable by direct comparison methods. Dynamic programming is a method of sequence alignment that can take gaps into account but that requires a manageable number of comparisons.

To understand how the method works, we must first recall what is meant by an alignment, using the two protein sequences. The two sequences will be written across the page, one under the other, the object being to bring as many amino acids as possible into register. In some regions, amino acids in one sequence will be placed directly below identical amino acids in the second. In other regions, this process may not be possible and nonidentical amino acids may have to be placed next to each other, or else gaps must be introduced into one of the sequences. Gaps are added to the alignment in a manner that increases the matching of identical or similar amino acids at subsequent portions in the alignment. Ideally, when two similar protein sequences are aligned, the alignment should have long regions of identical or related amino acid pairs and very few gaps. As the sequences become more distant, more mismatched amino acid pairs and gaps should appear.

The quality of the alignment between two sequences is calculated using a scoring system that favours the matching of related or identical amino acids and penalizes for poorly matched amino acids and gaps. To decide how to score these regions, information on the types of changes found in related protein sequences is needed. These changes may be expressed by the following probabilities: (i) that a particular amino acid pair is found in alignments of related proteins; (ii) that the same amino acid pair is aligned by chance in the sequences, given that some amino acids are abundant in proteins and others rare; and (iii) that the insertion of a gap of one or more residues in one of the sequences (the same as an insertion of the same length in the other sequence), thus forcing the alignment of each partner of the amino acid pair with another amino acid, would be a better choice. The ratio of the first two probabilities is usually provided in an amino acid substitution matrix. Each table entry gives the ratio of the observed frequency of substitution between each possible amino acid pair in related proteins to that expected by chance, given the frequencies of the amino acids in proteins. These ratios are called *odds scores*. The ratios are transformed to logarithms of odds scores, called *log odds* scores, so that scores of sequential pairs may be added to reflect the overall odds of a real to chance alignment of an alignment. Examples are the Dayhoff PAM250 and BLOSUM62 substitution matrices described below. These matrices contain positive and negative values, reflecting the likelihood of each amino acid substitution in related proteins. Using these tables, an alignment of a sequential set of amino acid pairs with no gaps receives an overall score that is the sum of the positive and negative log odds scores for each individual amino acid pair in the alignment. The higher this score, the more significant is the alignment, or the more it resembles alignments in related proteins. The score given for gaps in aligned sequences is negative, because such misaligned regions should be uncommon in sequences of related proteins. Such a score will reduce the score obtained from an adjacent, matching region upstream in the sequences. The score of the alignment, using values from the BLOSUM62 amino acid substitution matrix and a gap penalty score of –11 for a gap of length 1, is 26 (the sum of amino acid pair scores) –11 = 15. The value of –11 as a penalty for a gap of length 1 is used because this value is already known from experience to favour the alignment of similar regions when the BLOSUM62 comparison matrix is used. Choice of the gap penalty is discussed further below where a table giving suitable choices is presented. As shown in the example, the presence of the gap decreases significantly the overall score of the alignment.

Although one may be able to align the two short sequences by eye and to place the gap where shown, the dynamic programming algorithm will automatically place gaps in much longer sequence alignments so as to achieve the best possible alignment. The derivation of the dynamic programming algorithm is using the above alignment as an example. Consider building this alignment in steps, starting with an initial matching aligned pair of characters from the sequences (V/V) and then sequentially adding a new pair until the alignment is complete, at each stage choosing a pair from all the possible matches that provides the highest score for the alignment up to that point. If the full alignment finally reached on the left side has the highest possible or optimal score, then the old alignment from which it was derived (A) by addition of the aligned Y/Y pair must also have been optimal up to that point in the alignment. If this were incorrect, and a different preceding alignment other than A was the highest scoring one, then the alignment on the left would also not be the highest scoring alignment, and we started with that as a known condition. Similarly, alignment A must also have been derived from an optimal alignment (B) by addition of a C/C pair. In this manner, the alignment can be traced back sequentially to the first aligned pair that was also an optimal alignment. One concludes that the building of an optimal alignment in this stepwise fashion can provide an optimal alignment of the entire sequences.

The example also illustrates two of the three choices that can be made in adding to an alignment between two sequences: Match the next two characters in the next positions in each sequence, or match the next character to a gap in the upper sequence. The last possibility, not illustrated, is to add a gap to the lower sequence. This situation is analogous to performing a dot matrix analysis of the sequences, and of either continuing a diagonal or of shifting the diagonal sideway or downward to produce a gap in one of the sequences.

Formal Description of the Dynamic Programming Algorithm

The algorithm may be written in mathematical form. The diagram indicates the moves that are possible to reach a certain matrix position (i,j) starting from the previous row and column at position $(i-1, j-1)$ or from any position in the same row and column.

The following equation describes the algorithm. There are three paths in the scoring matrix for reaching a particular position, a diagonal move from position $i-1, j-1$ to position i, j with no gap penalties, or a move from any other position from column j or row i, with a gap penalty that depends on the size of the gap. For two sequences $a = a_1a_2 \ldots a_n$ and $b = b_1\ b_2 \ldots b_n$, where $S_{ij} = S(a_1a_2 \ldots a_i,\ b_1b_2 \ldots b_j)$ then

$$S_{ij} = \max \{ S_{i-1,j-1} + s(a_i b_j),\quad \max_{x \geq 1} (S_{i-x,j} - w_x),\quad \max_{y \geq 1} (S_{ij-y} - w_y) \} \quad \ldots(1)$$

where S_{ij} is the score at position i in sequence a and position j in sequence b, $s(a_ib_j)$ is the score for aligning the characters at positions i and j, w_x is the penalty for a gap of length x in sequence a, and w_y is the penalty for a gap of length y in sequence b. Note that S_{ij} is a type of running best score as the algorithm moves through every position in the matrix. Eventually, when all of the matrix positions (all S_{ij}) have been filled, the best score of the alignment will be found as the highest scoring position in the last row and column (for a global alignment), after correcting for any remaining gap penalties to align the sequence ends, if applicable. To determine an optimal alignment of the sequences from the scoring matrix, a second matrix called the *trace-back matrix* is used. The trace-back matrix keeps track of the positions in the scoring matrix that contributed to the highest overall score found. The sequence characters corresponding to these high scoring positions may align or may be next to a gap, depending on the information in the trace-back matrix.

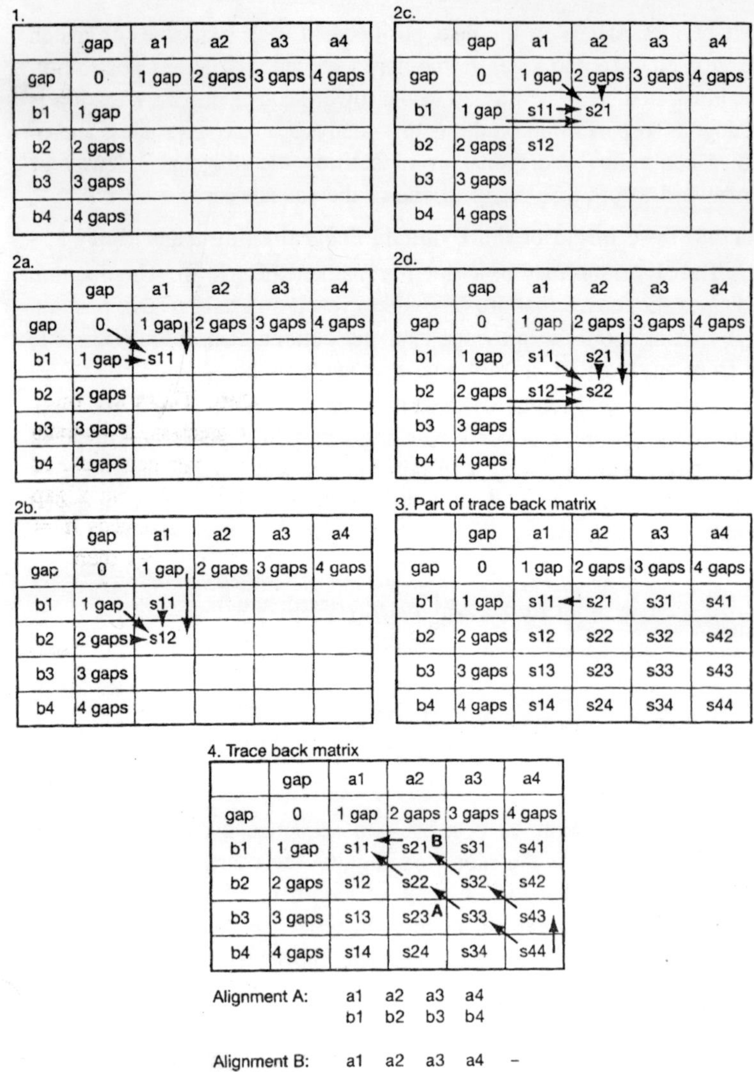

Fig. 4.5. Example of using the dynamic programming algorithm to align sequences a1 a2 a3 a4 and b1 b2 b3 b4.

Use of the dynamic programming method requires a scoring system for the comparison of symbol pairs (nucleotides for DNA sequences and amino acids for protein sequences), and a scheme for insertion/deletion (GAP) penalties. Once those parameters have been set, the resulting alignment for two sequences should always be the same.

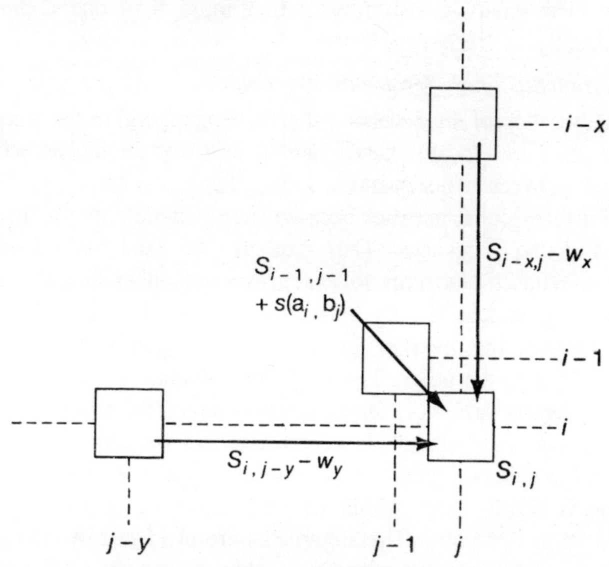

Fig. 4.6. Formal description of the dynamic programming algorithm.

Scoring matrices are described below. The most commonly used ones for protein sequence alignments are the log odds form of the PAM250 matrix and the BLOSUM62 matrix. However, a number of other choices are available.

Dynamic Programming can Provide Global or Local Sequence Alignments

Global alignment: Needleman-Wunsch algorithm

The dynamic programming method as described above gives a global alignment of sequences, as described by Needleman and Wunsch (1970), but was also proven mathematically and extended to include an improved scoring system by Smith and Waterman (1981a,b). The optimal score at each matrix position is calculated by adding the current match score to previously scored positions and subtracting gap penalties, if applicable. Each matrix position may have a positive or negative score, or 0. The Needleman-Wunsch algorithm will maximize the number of matches between the sequences along the entire length of the sequences. Gaps may also be present at the ends of sequences, in case there is extra sequence left over after the alignment. These end gaps are often, but not always, given a gap penalty. The effect of these penalties is illustrated below. An example of a global alignment of two short sequences calculated by hand using the algorithm is shown on the book

Web site. The example also reveals that more than one alignment may be equally as likely.

Local alignment: Smith-Waterman algorithm

A modification of the dynamic programming algorithm for sequence alignment provides a local sequence alignment giving the highest-scoring local match between two sequences. Local alignments are usually more meaningful than global matches because they include patterns that are conserved in the sequences. They can also be used instead of the Needleman-Wunsch algorithm to match two sequences that may have a matched region that is only a fraction of their lengths, that have different lengths, that overlap, or where one sequence is a fragment or subsequence of the other. The rules for calculating scoring matrix values are slightly different, the most important differences being (1) the scoring system must include negative scores for mismatches, and (2) when a dynamic programming scoring matrix value becomes negative, that value is set to zero, which has the effect of terminating any alignment up to that point. The alignments are produced by starting at the highest-scoring positions in the scoring matrix and following a trace path from those positions up to a box that scores zero. The mathematical formulation of the dynamic programming algorithm is revised to include a choice of zero as the minimum value at any matrix position. Fer two sequences $a = a_1a_2...a_n$ and $b = b_1 b_2...b_n$, where $H_{ij} = H(a_1a_2...a_i, b_1b_2..b_j)$, then

$$H_{ij} = \max \{H_{i-1,j-1} + s(a_i b_j), \max_{x\geq 1} (H_{i-x,j} - w_x), \max_{y\geq 1} (H_{i,j-y} - w_y), 0\} \quad ...(2)$$

where H_{ij} is the score at position i in sequence a and position j in sequence b, $s(a_ib_j)$ is the score for aligning the characters at positions i and j, w_x is the penalty for a gap of length x in sequence a, and w_y is the penalty for a gap of length y in sequence b.

Does a Local Alignment Program Always Produce a Local Alignment and a Global Alignment Program Always Produce a Global Alignment?

Although a computer program that is based on the above Smith-Waterman local alignment algorithm is used for producing an optimal alignment, this feature alone does not assure that a local alignment will be produced. The scoring matrix or match and mismatch scores and the gap penalties chosen also influence whether or not a local alignment is obtained. Similarly, a program based on the Needleman-Wunsch algorithm can also return a local alignment depending on the weighting of end gaps and on other scoring parameters. Often, one can simply inspect the alignment obtained to see how many gaps are present. It the matched regions are long and cover most of the sequences and obviously depend on the presence of many gaps, the alignment is global. A local alignment, on the other hand, will tend to be shorter and not include many gaps. However, these tests are quite subjective, and a more precise method of knowing whether a given program and set of scoring parameters will provide a local or global alignment is required. Looking ahead in the chapter for a moment, the best way of knowing is by looking at what happens when many random or completely unrelated sequences are aligned under the chosen conditions. As the length of the random sequences being aligned increases, the score of a global alignment will just increase proportionally. This is easy to see. Because a global alignment matches most of the sequence, and the negative mismatch score and gap penalties are deliberately chosen to be small in comparison to match scores in order to provide a long alignment, only matches count and the score has to be proportional to the length.

If using a scoring matrix, a matrix that gives on the average a positive score to each aligned position, combined with a small enough gap penalty to allow extension of the alignment through poorly matched regions, will give a global alignment. Conversely, for the local alignment, a negative mismatch score and gap penalties are chosen to balance the positive score of a match and to prevent the alignment growing into regions that do not match very well. The scoring matrix in this case will on the average give a negative value to the matched positions, and the gap penalty will be large enough to prevent gaps from extending the alignment. The local alignment score of random sequences does not increase proportionally to sequence length, because the positive score of matches is offset by the mismatch and penalty scores. In this case, it may be shown by theory and experiment that the score of local random alignments increases much more slowly, and proportionally to the logarithm of the product of the sequence lengths. It is this different behaviour of the alignment score of random sequences with length that distinguishes global and local alignments.

One may well ask, Does it really matter whether I use a sequence alignment program based on the global alignment algorithm or one based on the local alignment algorithm? The answer is that sometimes both methods will provide the same alignment with the same scoring system and sometimes they will not. The most reasonable approach is to use a program based on the appropriate algorithm for the analysis at hand, and then to choose the scoring system carefully. Small changes in the scoring system can abruptly change an alignment from a local to a global one. There are even examples in the bioinformatics literature where this feature of alignment scoring systems has been overlooked. The rest of this chapter is designed to provide a suitable guide for making the right choices.

Additional Development and Use of the Dynamic Programming Algorithm for Sequence Alignments

Use of distance scores for sequence alignment

As originally designed by Needleman and Wunsch and Smith and Waterman, the dynamic programming algorithm was used for sequence alignments scored on the basis of the similarity or identity of sequence characters. An alternative method is to score alignments based on differences between sequences and sequence characters; i.e., how many changes are required to change one sequence into another. Using this measure, the greater the distance between sequences, the greater the evolutionary time that has elapsed since the sequences diverged from a common ancestor. Hence, distance scores provide a more biologically natural way to compare sequences than do similarity scores. Using a distance scoring scheme, Sellers (1974, 1980) showed that the dynamic programming method could be used to provide an alignment that highlighted the evolutionary changes. Smith et al. (1981) and Smith and Waterman (1981b) showed that alignments based on a similarity scoring scheme could give a similar alignment.

Improvement in speed and memory requirement for the dynamic programming algorithm

The dynamic programming methods for sequence alignments originally required between $n \times m$ and $n \times m_2$ steps and storage in several matrices of size $n \times m$, where n is the length of the shorter sequence. On the book Web site, a series of improvements in this algorithm that reduced the number of steps and amount of memory required are described. These steps include: (1) a decreased number of steps in the alignment algorithm by Gotoh (1982); (2) a reduction in the amount of memory required to a linear function of sequence length; (3) ability to find near-optimal alignments and to align long sequences; and (4) ability to find the best-scoring alternative alignments that do not include alignments of the same sequence positions.

An alternative global alignment is found by giving the matrix position that begins with an alignment score of zero, and then all matrix positions that are affected by this change are recalculated. The next highest matrix score and the path leading to it provide an alternative alignment of the sequences that does not include the same sequence matches as were present in the original alignment. Alternative local alignments are found by a more complex algorithm (the SIM algorithm) that includes the improvements listed above.

Examples of Global and Local Alignments

An example of global and local alignments between two phage repressor proteins using the Genetics Computer Group (GCG) programs GAP and BESTFIT is shown. Note that the proteins are 58% similar in the carboxy-terminal domain, which is the region required for protein-protein interactions and a self-cleavage function that leads to phage induction. In these GCG implementations of the Needleman-Wunsch and Smith-Waterman algorithms, the alignments found in the carboxy-terminal domain are identical. However, the Smith-Waterman method (B) only reports the most alike regions, as expected by the focus on a local alignment strategy. In contrast, the Needleman-Wunsch method shows the entire alignment of the sequences but reports a lower score of similarity due to the longer alignment.

LALIGN is an implementation of the SIM algorithm for finding multiple unique (nonintersecting) alignments in DNA and protein sequences distributed in the FASTA package from W. Pearson. The program is also available on Web sites. Two features of these alignments are noteworthy: First, the highest-scoring alignment is similar to that found by the GAP program using a different amino acid substitution matrix and different gap penalties, with some minor variations in the more dissimilar regions and extension of the alignment farther into the amino-terminal domains. Second, by design, the alternative alignments never align the same amino acids and, in this example, the second and third alignments score much lower than the first one. These observations that strongly aligning regions are not significantly influenced by the scoring system, and that alternative high-scoring alignments are not possible, add convincing support that the initial alignment represents true similarity between these sequences.

USE OF SCORING MATRICES AND GAP PENALTIES IN SEQUENCE ALIGNMENTS

Amino Acid Substitution Matrices

Protein chemists discovered early on that certain amino acid substitutions commonly occur in related proteins from different species. Because the protein still functions with these substitutions, the substituted amino acids are compatible with protein structure and function. Often, these substitutions are to a chemically similar amino acid, but other changes also occur. Yet other substitutions are relatively rare. Knowing the types of changes that are most and least common in a large number of proteins can assist with predicting alignments for any set of protein sequences. If related protein sequences are quite similar, they are easy to align, and one can readily determine the single-step amino acid changes. If ancestor relationships among a group of proteins are assessed, the most likely amino acid changes that occurred during evolution can be predicted. Amino acid substitution matrices or symbol comparison tables, as they are sometimes called, are used for such purposes. Although the most common use of such tables is for comparison of protein sequences, other tables of nucleic acid symbols are also used for comparison of nucleic acid sequences in order to accommodate ambiguous nucleotide characters or models of expected sequence changes during different periods of evolutionary time that vary scoring of transitions and transversions.

In the amino acid substitution matrices, amino acids are listed both across the top of a matrix and down the side, and each matrix position is filled with a score that reflects how often one amino acid would have been paired with the other in an alignment of related protein sequences. The probability of changing amino acid A into B is always assumed to be identical to the reverse probability of changing B into A. This assumption is made because, for any two sequences, the ancestor amino acid in the phylogenetic tree is usually not known. Additionally, the likelihood of replacement should depend on the product of the frequency of occurrence of the two amino acids and on their chemical and physical similarities. A prediction of this model is that amino acid frequencies will not change over evolutionary time.

Dayhoff amino acid substitution matrices (percent accepted mutation or PAM matrices)

This family of matrices lists the likelihood of change from one amino acid to another in homologous protein sequences during evolution. There is presently no other type of scoring matrix that is based on such sound evolutionary principles as are these matrices. Even though they were originally based on a relatively small data set, the PAM matrices remain a useful tool for sequence alignment. Each matrix gives the changes expected for a given period of evolutionary time, evidenced by decreased sequence similarity as genes encoding the same protein diverge with increased evolutionary time. Thus, one matrix gives the changes expected in homologous proteins that have diverged only a small amount from each other in a relatively short period of time, so that they are still 50% or more similar. Another gives the changes expected of proteins that have diverged over a much longer period, leaving only 20% similarity. These predicted changes are used to produce optimal alignments between two protein sequences and to score the alignment. The assumption in this evolutionary model is that the amino acid substitutions observed over short periods of evolutionary history can be extrapolated to longer distances, The BLOSUM matrices are based on scoring substitutions found over a range of evolutionary periods and reveal that substitutions are not always as predicted by the PAM model.

In deriving the PAM matrices, each change in the current amino acid at a particular site is assumed to be independent of previous mutational events at that site. Thus, the probability of change of any amino acid *a* to amino acid *b* is the same, regardless of the previous changes at that site and also regardless of the position of amino acid *a* in a protein sequence. Amino acid substitutions in a protein sequence are thus viewed as a Markov model, characterized by a series of changes of state in a system such that a change from one state to another does not depend on the previous history of the state. Use of this model makes possible the extrapolation of amino acid substitutions observed over a relatively short period of evolutionary time to longer periods of evolutionary time.

To prepare the Dayhoff PAM matrices, amino acid substitutions that occur in a group of evolving proteins were estimated using 1572 changes in 71 groups of protein sequences that were at least 85% similar. Because these changes are observed in closely related proteins, they represent amino acid substitutions that do not significantly change the function of the protein. Hence they are called "accepted mutations," defined as amino acid changes "accepted" by natural selection. Similar sequences were first organized into a phylogenetic tree. The number of changes of each amino acid into every other amino acid was then counted. To make these numbers useful for sequence analysis, information on the relative amount of change for each amino acid was needed.

Relative mutabilities were evaluated by counting, in each group of related sequences, the number of changes of each amino acid and by dividing this number by a factor, called the exposure to mutation of the amino acid. This factor is the product of the frequency of occurrence of the amino acid in that group of sequences being analyzed and the total number of all amino acid changes that occurred in that group per 100 sites. This factor normalizes the data for variations in amino acid composition, mutation rate, and sequence length. The normalized frequencies were then summed for all sequence groups. By these scores, Asn, Ser, Asp, and Glu were the most mutable amino acids, and Cys and Trp were the least mutable.

The above amino acid exchange counts and mutability values were then used to generate a 20 × 20 mutation probability matrix representing all possible amino acid changes. Because amino acid change was modeled by a Markov model, the mutation at each site being independent of the previous mutations, the changes predicted for more distantly related proteins that have undergone *N* mutations could be calculated. By this model, the PAM 1 matrix could be multiplied by itself *N* times, to give transition matrices for comparing sequences with lower and lower levels of similarity due to separation of longer periods of evolutionary history. Thus, the commonly used PAM250 matrix represents a level of 250% of change expected in 2500 my. Although this amount of change seems very large, sequences at this level of divergence still have about 20% similarity. For example, alanine will be matched with alanine 13% of the time and with another amino acid 87% of the time.

The percentage of remaining similarity for any PAM matrix can be calculated by summing the percentages for amino acids not changing (Ala versus Ala, etc.) after multiplying each by the frequency of that amino acid pair in the database (e.g., 0.089 for Ala). The PAM120, PAM80, and PAM60 matrices should be used for aligning sequences that are 40%, 50%, and 60% similar, respectively. Simulations by George et al. (1990) have shown that, as predicted, the PAM250 matrix provides a better-scoring alignment than lower-numbered PAM matrices for distantly related proteins of 14-27%) similarity.

PAM matrices are usually converted into another form, called *log odds* matrices. The odds score represents the ratio of the chance of amino acid substitution by two different hypotheses—one that the change actually represents an authentic evolutionary variation at that site (the numerator), and the other that the change occurred because of random sequence variation of no biological significance (the denominator). Odds ratios are converted to logarithms to give log odds scores for convenience in multiplying odds scores of amino acid pairs in an alignment by adding the logarithms.

Table 4.1. Normalization probability scores for changing Phe to any other amino acids at PAM1 and PAM250 evolutionary distances

Amino acid change	*PAM1*	*PAM250*
Phe to Ala	0.0002	0.04
Phe to Arg	0.0001	0.01
Phe to Asn	0.0001	0.02
Phe to Asp	0.0000	0.01
Phe to Cys	0.0000	0.01
Phe to Gln	0.0000	0.01
Phe to Glu	0.0000	0.01
Phe to Gly	0.0001	0.03
Phe to His	0.0002	0.02
Phe to Ile	0.0007	0.05
Phe to Leu	0.0013	0.13
Phe to Lys	0.0000	0.02
Phe to Met	0.0001	0.02
Phe to Phe	0.9946	0.32
Phe to Pro	0.0001	0.02
Phe to Ser	0.0003	0.03
Phe to Thr	0.0001	0.03
Phe to Trp	0.0001	0.01
Phe to Tyr	0.0021	0.15
Phe to Val	0.0001	0.05
SUM	1.0000	1.00

At one time, the PAM250 scoring matrix was modified in an attempt to improve the alignment obtained. All scores for matching a particular amino acid were normalized to the same mean and standard deviation, and all amino acid identities were given the same score to provide an equal contribution for each amino acid in a sequence alignment. These modifications were included as the default matrices for the GCG sequence alignment programs in versions 8 and earlier and are optional in later versions. They are not recommended because they will not give an optimal alignment that is in accord with the evolutionary model.

Choosing the best PAM scoring matrices for detecting sequence similarity

The ability of PAM scoring matrices to distinguish statistically between chance and biologically meaningful alignments has been analyzed using a recently developed statistical theory for sequences that is discussed later in this chapter. As discussed above, each PAM matrix is designed to score alignments between sequences that have diverged by a particular degree of evolutionary distance. Altschul (1991) has examined how well the PAM matrices actually can distinguish proteins that have diverged to a greater or lesser extent, when these proteins are subjected to a local alignment.

Initially, when using a scoring matrix to produce an alignment, the amount of similarity between sequences may not be known. However, the ungapped alignment scores obtained are maximal when the correct PAM matrix, i.e., the one corresponding to the degree of similarity in the target sequences, is used. Altschul (1991) has also examined the ability of PAM matrices to provide a reliable enough indication of an ungapped local alignment score between sequences on an initial attempt of alignment, for sequence alignments, the PAM200 matrix is able to detect a significant ungapped alignment of 16-62 amino acids whose score is within 87% of the optimal one. Alternatively, several combinations, such as PAM80 and PAM250 or PAM120 and PAM350, can also in used. Altschul (1993) has also proposed using a single matrix and adjusting a statistical parameter in the scoring system to reach more distantly related sequences, but this change would primarily be for database searches.

Scoring matrices are also used in database searches for similar sequences. The optimal matrices for these searches have also been determined. It is important to remember that these predictions assume that the amino acid distributions in the set of protein families used to make the scoring matrix are representative of all families that are likely to be encountered. The original PAM matrices represent only a small number of families. Scoring matrices obtained more recently, such as the BLOSUM matrices, are based on a much larger number of protein families. BLOSUM matrices are not based on a PAM evolutionary model in which changes at large evolutionary distance are predicted by extrapolation of changes found at small distances. Matrix values are based on the observed frequency of change in a large set of diverse proteins. The BLOSUM scoring matrices (especially BLOSUM62) appear to capture more of the distant types of variations found in protein families.

In addition to the aforementioned differences among PAM scoring matrices for scoring alignments of more- or less-related proteins, the ability of each PAM matrix to discriminate real local alignments from chance alignments also varies. To calculate the ability of the entire matrix to discriminate related from unrelated sequences (H, the relative entropy), the score for each amino acid pair s_{ij} (in units of $\log_2$ called bits) is multiplied by the probability of occurrence of that pair in the original dataset, q_{ij}. This weighted score is then summed over all of the amino acid pairs to produce a score that represents the ability of the average amino acid pair in the matrix to discriminate actual from chance alignments.

$$H = \sum_{i=1}^{20} \sum_{j=1}^{i} q_{ij} \times s_{ij} \qquad \text{...(3)}$$

In information theory, this score is called the average mutual information content per pair, and the sum over all pairs is the relative entropy of the matrix (termed H). The relative entropy will be a small positive number. For the PAM250 matrix the number is +0.36, for PAM120, +0.98, and for PAM160, +0.70. In general, all other factors being equal, the higher the value of H for a scoring matrix, the more likely it is to be able to distinguish real from chance alignments.

Analysis of the Dayhoff model of protein evolution as used in PAM matrices

A outlined above, the Dayhoff model of protein evolution is a Markov process. In this mode each amino acid site in a protein can change at any time to any of the other 20 amino acid with probabilities given by the PAM table, and the changes that occur at each site are independent of the amino acids found at other sites in the protein and depend only on the current amino acid at the site. The assumptions that underlie the method of constructing the Dayhoff scoring matrix have been challenged. First, it is assumed that each amino acid position is equally mutable, whereas, in fact, sites vary considerably in their degree of mutability. Mutagenesis hot spots are well known in molecular genetics, and variations in mutability of different amino acid sites in proteins are well known. The more conserved amino acids in similar proteins from different species are ones that play an essential role in structure and function and the less conserved are in sites that can vary without having a significant effect on function. Thus, there are many factors that influence both the location and types of amino acid changes that occur in proteins. Wilbur (1985) has tested the Markov model of evolution and has shown that it can be valid it certain changes are made in the way that the PAM matrices are calculated.

A further criticism of the PAM scoring matrices is that they are not more useful for sequence alignment than simpler matrices, such as one based on a chemical grouping of amino acid side chains. Although alignment of related proteins is straightforward and quite independent of the symbol comparison scoring scheme, alignments of less-related proteins are much more speculative. These matrices and the BLOSUM matrices have been very useful for finding more distantly related sequences. There have been recent changes in the way that members of protein families are identified. Once a family has been identified, family-specific scoring matrices can be produced, and there is no point in using these general matrices. A scoring matrix representing a section of aligned sequences with no gaps, or a matrix representing a section of aligned sequences with matches, mismatches, and gaps, are the best tools to search for more family members.

Another criticism of the PAM matrix is that constructing phylogenetic relationships prior to scoring mutations has limitations, due to the difficulty of determining ancestral relationships among sequences. Early on in the Dayhoff analysis, the evolutionary trees were estimated by a voting scheme for the branches in the tree, each node being estimated by the most abundant amino acid in distal parts of the tree. Once available, the PAM matrices were used to estimate the evolutionary distance between proteins, given the amount of sequence similarity. Such data can be used to produce a tree based on evolutionary distances. This circular analysis of using alignments to score amino acid changes and then to use the matrices to produce new alignments has also been criticized. However, no method has yet been devised in any type of sequence analysis for completely circumventing this problem. Evidence that the values in the scoring matrix are insensitive to changes in the phylogenetic relationships has been provided.

Finally, the Dayhoff PAM matrices have been criticized because they are based on a small set of closely related proteins. The Dayhoff data set has been augmented to include the 1991 protein database. The ability of the Dayhoff matrices to identify homologous sequences has also been extensively compared to that of other scoring matrices.

Blocks amino acid substitution matrices (BLOSUM)

The BLOSUM62 substitution matrix is widely used for scoring protein sequence alignments. The matrix values are based on the observed amino acid substitutions in a large set of ~2000 conserved amino acid patterns, called *blocks*. These blocks have been found in a database of protein sequences representing more than 500 families of related proteins and act as signatures of these protein families. The BLOSUM matrices are thus based on an entirely different type of sequence analysis and a much larger data set than the Dayhoff PAM matrices.

These protein families were originally identified by Bairoch in the Prosite catalog. This catalog provides lists of proteins that are in the same family because they have a similar biochemical function. For each family, a pattern of amino acids that are characteristic of that function is provided. Henikoff and Henikoff (1991) examined each Prosite family for the presence of ungapped amino acid patterns (blocks) that were present in each family and that could be used to identify members of that family. To locate these patterns, the sequences of each protein family were searched for similar amino acid patterns by the MOTIF program of H. Smith, which can find patterns of the type aa1 d1 aa2 d2 aa3, where aa1 and aa2 are conserved amino acids and d1 and d2 are stretches of intervening sequence up to 24 amino acids long located in all sequences. These initial patterns were organized into larger ungapped patterns (blocks) between 3 and 60 amino acids long by the Henikoffs' PROTOMAT program. Because these blocks were present in all of the sequences in each family, they could be used to identify other members of the same family. Thus, the family collections were enlarged by searching the sequence databases for more proteins with these same conserved blocks.

The blocks that characterized each family provided a type of multiple sequence alignment for that family. The amino acid changes that were observed in each column of the alignment could then be counted. The types of substitutions were then scored for all aligned patterns in the database and used to prepare a scoring matrix, the BLOSUM matrix, indicating the frequency of each type of substitution. As previously described for the PAM matrices, BLOSUM matrix values were given as logarithms of odds scores of the ratio of the observed frequency of amino acid substitutions divided by the frequency expected by chance. This procedure of counting all of the amino acid changes in the blocks, however, can lead to an overrepresentation of amino acid substitutions that occur in the most closely related members of each family. To reduce this dominant contribution from the most alike sequences, these sequences were grouped together into one sequence before scoring the amino acid substitutions in the aligned blocks. The amino acid changes within these clustered sequences were then averaged. Patterns that were 60% identical were grouped together to make one substitution matrix called BLOSUM60, and those 80% alike to make another matrix called BLOSUM80, and so on. As with the PAM matrices, these matrices differ in the degree to which the more common amino acid pairs are scored relative to the less common pairs. Thus, when used for aligning protein sequences, they provide a greater or lesser distinction between the more common and less common amino acid pairs. The ability of these different BLOSUM matrices to distinguish real from chance alignments and to identify as many members as possible of a protein family has been determined.

Two types of analyses were performed: (1) an information content analysis of each matrix, as was described above for the PAM matrices, and (2) an actual comparison of the ability of each matrix to find members of the same families in a database search, discussed below. As the clustering percentage was increased, the ability of the resulting matrix to distinguish actual from chance alignments, defined as the relative entropy of the matrix or the average information content per residue pair, also increased. As clustering increased from 45% to 62%, the information content per residue increased from ~0.4 to 0.7 bits per residue, and was ~1.0 bits at 80% clustering. However, at the same time, the number of blocks that contributed information decreased by 25% between no clustering and 62% clustering. BLOSUM62 represents a balance between information content and data size.

Henikoff and Henikoff (1993) have prepared a set of interval BLOSUM matrices that represent the changes observed between more closely related or more distantly related representatives of each block. Rather than representing the changes observed in very alike sequences up to sequences that were *n*% alike to give a BLOSUM-*n* matrix, the new BLOSUM-*nm* matrix represented the changes observed in sequences that were between *n*% alike and *m*% alike. The idea behind these matrices was to have a set of matrices corresponding to amino acid changes in sequence blocks that are separated by different evolutionary distances.

Comparison of the PAM and BLOSUM amino acid substitution matrices

There are several important differences in the ways that the PAM and BLOSUM scoring matrices were derived, and these differences should be appreciated in order to interpret the results of protein sequence alignments obtained with these matrices. First, the PAM matrices are based on a mutational model of evolution that assumes amino acid changes occur as a Markov process, each amino acid change at a site being independent of previous changes at that site. Changes are scored in sequences that are 85% similar after predicting a phylogenetic history of the changes in each family. Thus, the PAM matrices are based on prediction of the first changes that occur as proteins diverge from a common ancestor during evolution of a protein family. Matrices that may be used to compare more distantly related proteins are then derived by extrapolation from these short-term changes, assuming that these more distant changes are a reflection of the short-term changes occurring over and over again. For each longer evolutionary interval, each amino acid can change to any other with the same frequency as observed in the short term. In contrast, the BLOSUM matrices are not based on an explicit evolutionary model. They are derived from considering all amino acid changes observed in an aligned region from a related family of proteins, regardless of the overall degree of similarity between the protein sequences. However, these proteins are known to be related biochemically and, hence, should share common ancestry. The evolutionary model implied in such a scheme is that the proteins in each family share a common origin, but closer versus distal relationships are ignored, as if they all were derived equally from the same ancestor, called a starburst model of protein evolution. Second, the PAM matrices are based on scoring all amino acid positions in related sequences, whereas the BLOSUM matrices are based on substitutions and conserved positions in blocks, which represent the most alike common regions in related sequences. Thus, the PAM model is designed to track the evolutionary origins of proteins, whereas the BLOSUM model is designed to find their conserved domains.

Other amino acid scoring matrices

In addition to the Dayhoff PAM, and related Gonnet et al. (1992), Benner et al. (1994), and Jones et al. (1992) matrices and the BLOSUM matrices, a number of other amino acid substitution matrices have been used for producing protein sequence alignments and several representative ones are listed in Table 4.2. These tables vary from a comparison of simple chemical properties of amino acids to a complex analysis of the substitutions found in secondary structural domains of proteins. Because most of these tables are designed to align proteins on the basis of some such feature of the amino acids, and not on an evolutionary model, they are not particularly suitable for evolutionary analysis. They can be very useful, however, for discovering structural and functional relationships, or family relationships among proteins. A sequence alignment program that uses a combination of these tables has been found to be particularly useful for detecting distant protein relationships.

Table 4.2. Criteria used in amino acid scoring matrices for sequence alignments

1. Simple identity, which scores only identical amino acids as a match and all others as a mismatch.
2. Genetic code changes, which score the minimum number of nucleotide changes to change a codon for one amino acid into a codon for another, due to Fitch (1966), and also with added information based on structural similarity of amino acid side chains. A similar matrix based on the assumption that genetic code is the only factor influencing amino acid substitutions has been produced.
3. Matrices based on chemical similarity of amino acid side chains, molecular volume, and polarity and hydrophobicity of amino acid side chains.
4. Amino acid substitutions in structurally aligned three-dimensional structures. A similar matrix was described by Henikoff and Henikoff (1993). Sander and Schneider (1991) prepared a similar matrix based on these same substitutions but augmented by substitutions found in proteins which are so similar to the structure-solved group that they undoubtedly have the same three-dimensional structure.
5. Gonnet et al. (1994) have prepared a 400 × 400 dipeptide substitution matrix for aligning proteins based on the possibility that amino acid substitutions at a particular site are influenced by neighbouring amino acids, and thus that the environment of an amino acid plays a role in protein evolution.
6. Jones et al. (1994) have prepared a scoring matrix specifically for transmembrane proteins. This matrix was prepared using an analysis similar to that used for preparing the original Dayhoff PAM matrices, and therefore provides an estimate of evolutionary distances among members of this class of proteins.

Nucleic Acid PAM Scoring Matrices

Just as amino acid scoring matrices have been used to score protein sequence alignments, nucleotide scoring matrices for scoring DNA sequence alignments have also been developed. The DNA matrix can incorporate ambiguous DNA symbols and information from mutational analysis, which reveals that transitions (substitutions between the purines A and G or between the pyrimidines C and T) are more probable than transversions (substitutions between purine to pyrimidine or pyrimidine to purine). These substitution matrices may be used to produce global or local alignments of DNA sequences.

States et al. (1991) have developed a series of nucleic acid PAM matrices based on a Markov transition model similar to that used to generate the Dayhoff PAM scoring matrices. Although designed to improve the sensitivity of similarity searches of sequence databases, these matrices also may be used to score nucleic acid alignments. The advantage of using these matrices is that they are based on a defined evolutionary model and that the statistical significance of alignment scores obtained by local alignment programs may be evaluated, as described later in this chapter.

To prepare these DNA PAM matrices, a PAM1 mutation matrix representing 99% sequence conservation and one PAM of evolutionary distance (1% mutations) was first calculated. For a model in which all mutations from any nucleotide to any other are equally likely, and in which the four nucleotides are present at equal frequencies, the four diagonal elements of the PAM1 matrix representing no change are 0.99 whereas the six other elements representing change are 0.00333. The values are chosen so that the sum of all possible changes for a given nucleotide in the PAM1 matrix is 1% ($3 \times 0.00333 = 0.00999$). For a biased mutation model in which a given transition is threefold more likely than a transversion, the off-diagonal matrix elements corresponding to the one possible transition for each nucleotide are 0.006 and those for the two possible transversions are 0.002, and the sum for each nucleotide is again 1% ($0.006 + 0.002 + 0.002 = 0.01$).

As with the amino acid matrices, the above matrix values are then used to produce log odds scoring matrices that represent the frequency of substitutions expected at increasing evolutionary distances. In terms of an alignment, the probability (s_{ij}) of obtaining a match between nucleotides i and j, divided by the random probability of aligning i and j, is given by

$$s_{ij} = \log(p_i M_{ij} / p_i p_j) \qquad \ldots(4)$$

where M_{ij} is the value in the mutation matrix, and p_i and p_j are the fractional composition of each nucleotide, assumed to be 0.25. The base of the logarithm can be any value, corresponding to multiplying

Table 4.3. Nucleotide mutation matrix for an evolutionary distance of 1 PAM, which corresponds to a probability of a change at each nucleotide position of 1%

A. Model of uniform mutation rate among nucleotides

	A	G	T	C
A	0.99			
G	0.00333	0.99		
T	0.00333	0.00333	0.99	
C	0.00333	0.00333	0.00333	0.99

B. Model of threefold higher transitions than transversions

	A	G	T	C
A	0.99			
G	0.006	0.99		
T	0.002	0.002	0.99	
C	0.002	0.002	0.006	0.99

every value in the matrix by the same constant. With such scaling variations, the ability of the matrix to distinguish among significant and chance alignments will not be altered. The resulting tables with s_{ij} expressed in units of bits (logarithm to the base 2) and rounded off to the nearest whole integer.

Table 4.4. Nucleotide substitution matrix at 1 PAM of evolutionary distance

A. Model of uniform mutation rates among nucleotides

	A	G	T	C
A	2			
G	–6	2		
T	–6	–6	2	
C	–6	–6	–6	2

B. Model of threefold higher transitions than transversions

	A	G	T	C
A	2			
G	–5	2		
T	–7	–7	2	
C	–7	–7	–5	2

From these PAM1 matrices, additional log odds matrices at an evolutionary distance of *n* PAMs may be obtained by multiplying the PAM1 matrix by itself *n* times. The ability of each matrix to distinguish real from random nucleotide matches in an alignment, designated *H*, measured in bit units ($\log_2$) can be calculated using the equation

$$H = \sum_{i,j} p_i p_j s_{ij} 2^{s_{ij}} \qquad \ldots(5)$$

where the s_{ij} scores are also expressed in bit units. In Table 4.5 are shown the log odds values of the match and mismatch scores for PAM matrices at increasing evolutionary distances, assuming a uniform rate of mutation among all nucleotides. Also shown is the percentage of nucleotides that will be changed at that distance. The identity score will be 100 minus this value. This percentage is not as great as the PAM score due to expected back-mutation over longer time periods. Also shown are the H scores of the matrices at each PAM value.

Table 4.5. Properties of nucleic acid substitution matrices assuming a uniform rate of mutation among nucleotides

PAM distance	*Percentage difference*	*Match score (bits)*	*Mismatch score (bits)*	*Average information per position (bits)*
10	9.4	1.86	–3.00	1.40
25	21.3	1.66	–1.82	0.92
50	36.5	1.34	–1.04	0.47
100	55.2	0.84	–0.44	0.13
125	60.8	0.65	–0.30	0.07

The following points may be made:

1. If comparing sequences that are quite similar, it is better to use a lower scoring matrix because the information content of the small PAM matrices is relatively higher. As discussed earlier for lower-numbered Dayhoff PAM matrices for more-alike protein sequences, a more optimal alignment will be obtained.
2. As the PAM distance increases, the mismatch scores in the biased mutational model become positive and appear as conservative substitutions. Thus, the bias model can provide considerably more information than the uniform mutation model when aligning sequences that are distantly related (>30% different) and may be used for this purpose.

3. The scoring matrices at large evolutionary distances provide very little information per aligned nucleotide pair. When sequences have so little similarity, a much longer alignment is necessary to be significant.

As with amino acid scoring matrices, the average information content shown is only achieved by using the scoring matrix that matches the percentage difference between the sequences. For example, for sequences that are 21% different (79% identical), the matrix at 25 PAM distance should be used. One cannot know ahead of time what the percentage similarity or difference between two sequences actually is until an alignment is done, thus a trial alignment must first be done. States et al. (1991) have calculated how efficient a given scoring matrix is at achieving the highest possible score in aligning two sequences that vary in their levels of similarity. Once the initial similarity score has been obtained with these matrices, a more representative score can be obtained by using another PAM matrix designed specifically for sequences at that level of similarity.

Gap Penalties

The inclusion of gaps and gap penalties is necessary in order to obtain the best possible alignment between two sequences. A gap opening penalty for any gap (*g*) and a gap extension penalty for each element in the gap (*r*) is most often used, to give a total gap score w_x, according to the equation

$$w_x = g + rx \qquad \text{...(6)}$$

where x is the length of the gap. Note that in some formulations of the gap penalty, the equation $w_x = g + r(x - 1)$ is used. Thus, the gap extension penalty is not added to the gap opening penalty until the gap size is 2. Although this difference does not affect the alignment obtained, one needs to distinguish which method is being used by a particular computer program if the correct results are to be obtained. In the former case, the penalty for a gap of size 1 is $g + x$, whereas in the latter case this value is g· The values for these penalties have to be chosen to balance the scores in the scoring matrix that is used. Thus, the Dayhoff log odds matrix at PAM250 is expressed in units of $\log_{10}$, which is approximately 1/3 bits, but if this matrix were converted to 1/2 bits, the same gap penalties would no longer be appropriate.

If too high a gap penalty is used relative to the range of scores in the substitution matrix, gaps will never appear in the alignment. Conversely, if the gap penalty is too low compared to the matrix scores, gaps will appear everywhere in the alignment in order to align as many of the same characters as possible, fortunately, most alignment programs will suggest gap penalties that are appropriate for a given scoring matrix in most situations. In the GCG and FASTA program suites, the scoring matrix itself is formatted in a way that includes default gap penalties. Examples of the values of *g* and *r* used by various alignment programs are shown on the book Web site. When deciding gap penalties for local alignment programs, another consideration is that the penalties should be large enough to provide a local alignment of the sequences. Altschul and Gish (1996) and Pearson (1996, 1998) have found that use of appropriate gap penalties will provide an improved local alignment based on statistical analysis.

Mathematician Peter Sellers (1974) showed that if sequence alignment was formulated in terms of distances instead of similarity between sequences, a biologically more appealing interpretation of gaps is possible. The distance is the number of changes that must be made to convert one sequence into the other and represents the number of mutations that will have occurred following separation of the genes during evolution; the greater the distance, the more distantly related are the sequences in evolution. In this case, substitution produces a positive score of 1. Notice that the distance score plus the similarity score for an alignment is equal to 1. Sellers proved that this distance formulation of sequence alignment has a desirable mathematical property that also makes evolutionary sense. If three sequences, *a*, *b*, and *c*, are compared using the above scoring scheme, the distance score as defined above is described as a metric that satisfies the triangle inequality relationship

$$d(a,b) + d(b,c) \geq d(a,c) \qquad \ldots(7)$$

where $d(a,b)$ is the distance between sequences *a* and *b*, and likewise for the other two *d* values. Expressed another way, if the three possible distances between three sequences are obtained, then the distance between any first pair plus that for any second pair cannot underscore the third pair. Violating this rule would not be consistent with the expected evolutionary origin of the sequences. To satisfy the metric requirement, the scoring of individual matches, mismatches, and gaps must be such that in an alignment of two identical sequences *a* and a', $d(a,a')$ must equal 0 and for two totally different sequences *b* and b', $d(b,b')$ must equal 1. For any other two sequences *a* and *b*, $d(a,b) = d(b,a)$. Hence, it is important that the distance score for changing one sequence character into a second is the same as the converse score for changing the second into the first, if the distance score of the alignment is to remain a metric and to make evolutionary sense. The above relationships were shown by Sellers to be true for gaps of length 1 in a sequence alignment. He also showed that the smallest number of steps required to change one sequence into the other could be calculated by the dynamic programming algorithm. The method was similar to that discussed above for the Needleman-Wunsch global and Smith-Waterman local alignments, except that these former methods found the maximum similarity between two sequences, as opposed to the minimum distance found by the Sellers analysis.

Subsequently, Smith et al. (1981) and Smith and Waterman (1981a,b) showed that gaps of any length could also be included in an alignment and still provide a distance metric for the alignment score. In this formulation, the gap penalty was required to increase as a function of the gap length. The argument was made that a single mutational event involving a single gap of *n* residues should be more likely to have occurred than *n* single gaps. Thus, to increase the likelihood of such gaps of length > 1 being found, the penalty for a gap of length *n* was made smaller than the score for *n* individual gaps. The simplest way of implementing this feature of the gap penalty was to have the gap score w_x be a linear function of gap length by consisting of two parts, a larger gap opening penalty (*g*) and a smaller gap extension penalty (*r*) for each extra position in the gap, or $w_x = g + rx$, where *x* is the length of the gap, as described above. This type of gap penalty's referred to as an affine gap penalty in the literature. Any other formula for scoring gap penalties should also work, provided that the score increases with length of the gap but that the score is less than *x* individual gaps. Scoring of gaps by the above linear function of gap length has now become widely used in sequence alignment. However, more complex gap penalty functions have been used.

Penalties for gaps at the ends of alignments

Sequence alignments are often produced that include gaps opposite nonmatching characters at the ends of an alignment. These gaps may be given the same penalty score as gaps inside of the alignment or, alternatively, they may not be given any penalty score. End gaps were an important component in the mathematical formulation of both the similarity and distance methods of sequence alignment for producing both global and local alignments. Failure to include them in distance calculations can result in a failure to obtain distance scores that make evolutionary sense. Without scoring end alignments, gaps may be liberally placed at the ends of alignments by the dynamic programming algorithm to increase the matching of internal characters, as opposed to including these gaps as a part of the overall alignment.

If comparing sequences that are homologous and of about the same length, it makes a great deal of sense to include end gap penalties to achieve the best overall alignment. For sequences that are of unknown homology or of different lengths, it may be better to use an alignment that does not include end gap penalties. If one sequence is expected to be contained within the other, it is reasonable to include end gap penalties only for the shorter sequence. However, for any test alignment, these end penalties should be included in at least one alignment to assure that they do not have an effect. It is also important to use alignment programs that include them as an option.

Parametric sequence alignments

Computer methods that find a range of possible alignments in response to varying the scoring system used for matches, mismatches, and gaps, called *parametric* sequence comparisons, have been developed. There is also an effort to use scores such that the results of global and local types of sequence alignments provide consistent results. For example, if two sequences are similar along their entire lengths, both global and local methods should provide the same alignment. The program Xparal, which can perform this type of analysis, is available from http://theory.cs.ucdavis.edu/~stevenk. The program runs on a UNIX environment under X-Windows. When provided with two sequences and some of the alignment parameters, such as gap score, the program displays graphically the types of possible alignments when the remaining parameters are varied. Another sequence alignment program that performs parametric sequence alignment is the Bayes block aligner.

Effects of varying mismatched gap penalties on local alignment scores

Vingron and Waterman (1994) have reviewed the effect of varying the parameters of the scoring system on the alignment of random DNA and protein sequences. To simplify the number of parameters, a constant penalty for any size gap was used. If a very high mismatch penalty is used relative to a positive score for a match, with zero gap penalty, the local alignment of these sequences will not include any gaps and is defined as the longest common subsequence. The global alignment with the same scoring parameters will have no mismatches but will have many gaps so placed as to maximize the matches, and the score will be positive. In this case, the score of the local alignment of the sequences is predicted to increase linearly with the length of the sequences being compared.

Another case of varying alignment is penalizing gaps heavily. Then the best scoring local alignment between the sequences will be one that optimizes the score between matches and mismatches, without any gaps. If both mismatches and gaps are heavily penalized, the resulting alignment will also be a local alignment that contains the longest region of exact matches. In the above two cases, the alignment score of the highest-scoring local alignment will increase as the logarithm of the length of the sequences. Under these same conditions, the score of the corresponding global alignment between the sequences will be negative. The transition between a linear and logarithmic dependence of the local similarity score on sequence length occurs when the score of the corresponding global alignment is zero. When both the mismatch and gap penalties are varied between zero and a high negative score, the number of possible alignments of random DNA sequences is very large.

Three general conclusions can be drawn from this theoretical study of random sequence alignments: (1) Use of high mismatch and gap penalties that are greater than a match score will find local alignments, of which there are relatively few in number; (2) when the penalty for a mismatch is greater than twice the score for a match, the gap penalty becomes the decisive parameter in the alignment; and (3) for a mismatch penalty less than twice the score of a gap and a wide range of gap penalties, there are a large number of possible alignments that depend on both the mismatch and gap penalty scores.

Distinguishing local from global alignments has an important practical application. A local alignment is rarely produced between random sequences. Accordingly, the significance of a local alignment between real sequences may be readily calculated, as described below. In contrast, the significance of a global alignment is difficult to determine since a global alignment is readily produced between random sequences.

Optimal Combinations of Scoring Matrices and Gap Penalties for Finding Related Proteins

The usefulness of combinations of scoring matrices and gap penalties for identifying related proteins, including distantly related ones, has been compared. The method generally used is to start with a database of protein sequences organized into families, either based on sequence similarity or structural similarity. A member of a family is then selected and used as a query sequence in a search of the entire database from which the sequence came, using a database similarity search method (FASTA, BLAST, SSEARCH). These methods basically use the dynamic programming algorithm and a choice of scoring matrix and gap penalties to produce alignment scores.

In summary, the following general observations have been made: (1) Some scoring matrices are superior to others at finding related proteins based on either sequence or structure. For example, matrices prepared by examining the full range of amino acid substitutions in families of related proteins, such as the BLOSUM62 matrix, perform better than matrices based on variations in closely related proteins that are extrapolated to produce matrices for more distantly related sequences, such as the Dayhoff PAM250 matrix. (2) Gap penalties that for a given scoring matrix are adjusted to produce a local alignment are the most suitable. (3) To identify related sequences, the significance of the alignment scores should be estimated, as described in the following section.

These methods provide the means to demonstrate sequence similarity in even the most distantly related proteins. For closely related proteins, a PAM-type scoring matrix that matches the evolutionary separation of the sequences may provide a higher-scoring alignment. Another set of studies has suggested that a global alignment algorithm in combination with scoring matrices that have all positive values and suitable gap penalties can be used to align proteins that have limited sequence similarity (i.e., 25% identity) but that have similar structure.

Assessing the Significance of Sequence Alignments

One of the most important recent advances in sequence analysis is the development of methods to assess the significance of an alignment between DNA or protein sequences. For sequences that are quite similar, such as two proteins that are clearly in the same family, such an analysis is not necessary. A significance question arises when comparing two sequences that are not so clearly similar but are shown to align in a promising way. In such a case, a significance test can help the biologist to decide whether an alignment found by the computer program is one that would be expected between related sequences or would just as likely be found if the sequences were not related. The significance test is also needed to evaluate the results of a database search for sequences that are similar to a sequence by the BLAST and FASTA programs. The test will be applied to every sequence matched so that the most significant matches are reported. Finally, a significance test can also help to identify regions in a single sequence that have an unusual composition suggestive of an interesting function. Our present purpose is to examine the significance of sequence alignment scores obtained by the dynamic programming method.

Originally, the significance of sequence alignment scores was evaluated on the basis of the assumption that alignment scores followed a normal statistical distribution. If sequences are randomly generated in a computer by a Monte Carlo or sequence shuffling method, as in generating a sequence by picking marbles representing four bases or 20 amino acids out of a bag (the number of each type is proportional to the frequency found in sequences), the distribution may look normal at first glance. However, further analysis of the alignment scores of random sequences will reveal that the scores follow a different distribution than the normal distribution called the Gumbel extreme value distribution. In this section, we review some of the earlier methods used for assessing the significance of alignments, then describe the extreme value distribution, and finally discuss some useful programs for this type of analysis with some illustrative examples.

The statistical analysis of alignment scores is much better understood for local alignments than for global alignments. Recall that the Smith-Waterman alignment algorithm and the scoring system used to produce a local alignment are designed to reveal regions of closely matching sequence with a positive alignment score. In random or unrelated sequence alignments, these regions are rarely found. Hence, their presence in real sequence alignments is significant, and the probability of their occurring by chance alignment of unrelated sequences can be readily calculated. The significance of the scores of global alignments, on the other hand, is more difficult to determine. Using the Needleman-Wunsch algorithm and a suitable scoring system, there are many ways to produce a global alignment between any pair of sequences, and the scores of many different alignments may be quite similar. When random or unrelated sequences are compared using a global alignment method, they can have very high scores, reflecting the tendency of the global algorithm to match as many characters as possible. Thus, assessment of the statistical significance of a global alignment is a much more difficult task. Rather than being used as a strict test for sequence homology, a global alignment is more appropriately used to align sequences that are of approximately the same length and already known to be related. The method will conveniently show which sequence characters align. One can then use this information to perform other types of analyses, such as structural modeling or an evolutionary analysis.

Significance of Global Alignments

In general, global alignment programs use the Needleman-Wunsch alignment algorithm and a scoring system that scores the average match of an aligned nucleotide or amino acid pair as a positive number. Hence, the score of the alignment of random or unrelated sequences grows proportionally to the length of the sequences. In addition, there are many possible different global alignments depending on the scoring system chosen, and small changes in the scoring system can produce a different alignment. Thus, finding the best global alignment and knowing how to assess its significance is not a simple task, as reflected by the absence of studies in the literature.

Waterman (1989) provided a set of means and standard deviations of global alignment scores between random DNA sequences, using mismatch and gap penalties that produce a linear increase in score with sequence length, a distinguishing feature of global alignments. However, these values are of limited use because they are based on a simple gap scoring system. Abagyan and Batalov (1997) suggested that global alignment scores between unrelated protein sequences followed the extreme value distribution, similar to local alignment scores. However, since the scoring system that they used favoured local alignments, these alignments they produced may not be global but local. Unfortunately, there is no equivalent theory on which to base an analysis of global alignment scores as there is for local alignment scores. For zero mismatch and gap penalties, which is the most extreme condition for a global alignment giving the longest subsequence common to two sequences, the score between two random or unrelated sequences P is proportional to sequence length n, such that $P = cn$, but it has not proven possible to calculate the proportionality constant c.

To evaluate the significance of a Needleman-Wunsch global alignment score, Dayhoff (1978) and Dayhoff et al. (1983) evaluated Needleman-Wunsch alignment scores for a large number of randomized and unrelated but real protein sequences, using their log odds scoring matrix at 250 PAMs and a constant gap penalty. The distribution of the resulting random scores matched a normal distribution. On the basis of this analysis, the significance of an alignment score between two apparently related sequences A and B was determined by obtaining a mean and standard deviation of the alignment scores of 100 random permutations or shufflings of A with 100 of B, conserving the length and amino acid composition of each. If the score between A and B is significant, the authors specify that the real score should be at least 3-5 standard deviations greater than the mean of the random scores. This level of significance means that the probability that two unrelated sequences would give such a high score is 1.35×10^{-3} (3 S.D.S) and 2.87×10^{-6} (5 S.D.S). In evaluating an alignment, two parameters were varied to maximize the alignment score: First, a constant called the matrix bias was added to each value in the scoring matrix and, second, the gap penalty was varied. The statistical analysis was then performed after the score between A and B had been maximized. Recall that the log odds PAM250 matrix values vary from –7 to 17 in units of 1/3 bits. The bias varied from 2 to 20 and had the effect of increasing the score by the bias times the number of alignment positions where one amino acid is matched to another. As a result, the alignment frequently decreases in length because there are fewer gaps, assuming the gap penalty is not also changed. It was these optimized alignments on which the significance test was performed. Feng et al. (1985) used the same method to compare the significance of alignment scores obtained by using different scoring matrices. They used 25-100 pairs of randomized sequences for each test of an alignment.

There are several potential problems with this approach, some of which apply to other methods as well. First, the method is expensive in terms of the number of computational steps, which increase at least as much as the square of sequence length because many Needleman-Wunsch alignments must be done. However, this problem is much reduced with the faster computers and more efficient algorithms of today. Second, if the amino acid composition is unusual, and it there is a region of low complexity (for example, many occurrences of one or two amino acids), the analysis will be oversimplified. Third, when natural sequences were compared more closely, the patterns found did not conform to a random set of the basic building blocks of sequences but rather to a random set of sequence segments that were varying. Consider use of the 26-letter alphabet in English sentences. Alphabet letters do not appear in any random order in these sentences but rather in a vocabulary of meaningful words. What happens if sentences, which are made up of words, are compared? On the one hand, if just the alphabet composition of many sentences is compared, not much variation is seen. On the other hand, if words are compared, much greater variation is found because there are many more words than alphabet characters. If random sequences are produced from segments of sequences, rather than from individual residues, more variation is observed, more like that observed when unrelated natural sequences are compared. The increased variation found among natural sequences is not surprising when one thinks of DNA and proteins as sources of information. For example, protein-encoding regions of DNA sequences are constrained by the genetic code and by amino acid patterns that produce functional domains in proteins.

Lipman et al. (1984) analyzed the distribution of scores among 100 vertebrate nucleic acid sequences and compared these scores with randomized sequences prepared in different ways. When the randomized sequences were prepared by shuffling the sequence to conserve base composition, as was done by Dayhoff and others, the standard deviation was approximately one-third less than the distribution of scores of the natural sequences. Thus, natural sequences are more variable than randomized ones, and using such randomized sequences for a significance test may lead to an overestimation of the significance. If, instead, the random sequences were prepared in a way that maintained the local base composition by producing them from overlapping fragments of sequence, the distribution of scores has a higher standard deviation that is closer to the distribution of the natural sequences. The conclusion is that the presence of conserved local patterns can influence the score in statistical tests such that an alignment can appear to be more significant than it actually is. Although this study was done using the Smith-Waterman algorithm with nucleic acids, the same cautionary note applies for other types of alignments. The final problem with the above methods is that the correct statistical model for alignment scores was not used. However, these earlier types of statistical analysis methods set the stage for later ones.

The GCG alignment programs have a RANDOMIZATION option, which shuffles the second sequence and calculates similarity scores between the unshuffled sequence and each of the shuffled copies. If the new similarity scores are significantly smaller than the real alignment score, the alignment is considered significant. This analysis is only useful for providing a rough approximation of the significance of an alignment score and can easily be misleading.

Dayhoff (1978) and Dayhoff et al. (1983) devised a second method for testing the relatedness of two protein sequences that can accommodate some local variation. This method is useful for finding repeated regions within a sequence, similar regions that are in a different order in two sequences, or a small conserved region such as an active site. As used in a computer program called RELATE, all possible segments of a given length of one sequence are compared with all segments of the same length from another. An alignment score using a scoring matrix is obtained for each comparison to give a score distribution among all of the segments. A segment comparison score in standard deviation units is calculated as the difference between the value for real sequences minus the average value for random sequences divided by the standard deviation of the scores from the random sequences. A version of the program RELATE that runs on many computer platforms is included with the FASTA distribution package by W. Pearson. This program also calculates a distribution based on the normal distribution, thus it provides only an approximate indication of the significance of an alignment.

Modeling a Random DNA Sequence Alignment

The above types of analyses assume that alignment scores between random sequences follow a normal distribution that can be used to test the significance of a score between two test sequences, for a number of reasons, mathematicians were concerned that this statistical model might not be correct. Let's start by creating two aligned random DNA sequences by drawing pairs of marbles from a large bag filled with four kinds of labeled marbles. The marbles are in equal proportions and labeled A, T, G, and C to represent an assumed equal representation of the four nucleotides in DNA. Now consider the probability of removing 10 identical pairs representing 10 columns in an alignment between two random sequences. The probability of removing an identical pair (an A and another A) is $1/4 \times 1/4$, but there are 4 possible identical pairs (A/A, C/C, G/G, and T/T), so that the probability of removing any identical pair is $4 \times 1/4 \times 1/4 = 1/4$ and that for removing 6 identical pairs is $(1/4)^6 = 2.4 \times 10^{-4}$. The probability of drawing a mismatched pair is $1 - 1/4 = 3/4$, and that of drawing 6/6 mismatched pairs $(3/4)^6 = 0.178$. Most random alignments produced in this manner will have a mixture of a few matches and many mismatches.

The calculations are a little more complex if the four nucleotides are not equally represented, but the results will he approximately the same. The probability of drawing the same pair is p, where $p = p_A^2 + p_C^2 + p_G^2 + p_r^2$, where p_x is the proportion of nucleotide X. p is an important parameter to remember for the discussion below. An even more complicated situation is when the two random sequences to align have different nucleotide distributions. One way would be to use an average p for the two sequences. This example illustrates the difficulty of modeling sequence alignments between two different organisms that have a different base composition.

The above model is not suitable for predicting the number of sequentially matched positions between random sequences of a given length. To estimate this number, a DNA sequence alignment may also be modeled by coin-tossing experiments. Random alignments will normally comprise mixtures of matches and mismatches, just as a series of coin tosses will produce a mixture of heads and tails. The chance of producing a series of matches in a sequence alignment with no mismatches is similar to the chance of tossing a coin and coming up with a series of only heads. The numbers of interest are the highest possible score that can be obtained and the probability of obtaining such a score in a certain number of trials. In such models, coins are usually considered to be "fair" in that the probability of a head is equal to that of a tail. The coin in this example has a certain probability p of scoring a head (H) and $q = 1 - p$ of scoring a tail (T). The longest run of heads R has been shown by Erdos and Renyi to be given by $\log_{1/p}(n)$. If $p = 0.5$ as for a normal coin, then the base of the logarithm is $1/p = 2$. For the example of $n = 100$ tosses, then R $= \log_2 100 = \log_e 100/\log_e 2 = 4.605/0.693 = 6.65$.

To use the coin model, an alignment of two random sequences $a = a_1, a_2, a_3$---a_n and $b = b_1, b_2, b_3$---b_n, each of the same length n is converted to a series of heads and tails. If $a_i = b_i$ then the equivalent toss result is an H, otherwise the result is a T. The following example illustrates the conversion of an alignment to a series of H and T tosses.

$a_1\ a_2\ a_3$ ---a_n ---> HTH ---

$b_1\ b_2\ b_3$ ---b_n were $a_1 = b_1$ and $a_3 = b_3$ only ...(8)

The longest run of matches in the alignment is now equivalent to the longest run of heads in the coin-tossing sequence, and it should be possible to use the Erdos and Renyi law to predict the longest run of matches. This score, however, only applies to one particular alignment of random sequences, such as generated above by the marble draw. In performing a sequence alignment, two sequences are in effect shifted back and forth with respect to each other to find regions that can be aligned. In addition, the sequences may be of different lengths. If two random sequences of length m and n are aligned in this same manner, the same law still applies but the length of the predicted match is $\log_{1/p}(mn)$. If $m = n$, the longest run of matches is doubled. Thus, for

DNA sequences of length 100 and $p = 0.25$ (equal representation of each nucleotide), the longest expected run of matches is $2 \times \log_{1/p}(n) = 2 \times \log_4 100 = 2 \times \log_e 100/\log_e 4 = 2 \times 4.605/1.386 = 6.65$, the same number as in the coin-tossing experiment. This number corresponds to the longest subalignment that can be expected between two random sequences of this length and composition.

A more precise formula for the expectation value or mean of the longest match M and its variance has been derived.

$$E(M) \simeq \log_{1/p}(mn) + \log_{1/p}(q) + \gamma \log(e) - 1/2 \qquad \ldots(9)$$

$$\text{Var}\,[M(n,m)] \simeq [\pi \log_{1/p}(e)]^2/6 + 1/12 \qquad \ldots(10)$$

where $\gamma = 0.577$ is Euler's number and $q = 1 - p$. Note that Equation 9 can be simplified

$$E(M) \simeq \log_{1/p}(\mathrm{K}mn) \qquad \ldots(11)$$

where K is a constant that depends on the base composition.

Equation 11 also applies when there are k mismatches in the alignment, except that another term $= k \log_{1/p} \log_{1/p}(qmn)$ appears in the equation. K, the constant in Equation 11, depends on k. The log log term is small and can be replaced by a constant, and simulations also suggest that it is not important. Altschul and Gish (1996) have found a better match to Equation 11 when the length of each sequence is reduced by the expected length of a match. In the example given above with two sequences of length 100, the expected length of a match was 6.65. As the sequences slide align each other, it is not possible to have overlaps on the ends that are shorter than 7 because there is not enough sequence remaining. Hence, the effective length of the sequences is $100 - 7 = 93$. This correction is also used for the calculation of statistical significance by the BLAST algorithm.

Equation 11 is fundamentally important for calculating the statistical significance of alignment scores. Basically, it states that as the lengths of random or unrelated sequences increase, the mean of the highest possible local alignment scores will be proportional to the logarithm of the product of the sequence lengths, or twice the logarithm of the sequence length if the lengths are equal (since $\log(nn) = 2 \log n$). Equation 10 also predicts a constant variance among scores of random or unrelated sequences, and this prediction is also borne out by experiment. It is important to emphasize once again that this relationship depends on the use of scoring parameters appropriate for a local alignment algorithm, such as 1 for a match and –0.9 for a mismatch, or a scoring matrix that scores the average aligned position as negative, and also upon the use of sufficiently large gap penalties.

This type of scoring system gives rise to positive scoring regions only rarely. The significance of these scores can then be estimated as described herein.

Another way of describing the result in Equation 11 uses a different parameter, λ, where $\lambda = \log_e(1/p)$

$$E\ (M) = [\log_e\ (Kmn)]/\lambda \qquad ...(12)$$

Recall that p is the probability of a match between the same two characters, given above as 1/4 for matching a random pair of DNA bases, assuming equal representation of each base in the sequences. p may also be calculated as the probability of a match averaged over scoring matrix and sequence composition values. Instead, it is λ that is more commonly used with scoring matrix values. The calculation of λ and also of K is described below and in more detail on the book Web site.

It is more useful in sequence analysis to use alignment scores instead of lengths for comparing alignments. The expected or mean alignment length between two random sequences given by Equations 11 and 12 can be easily converted to an alignment score just by using match and mismatch or scoring matrix values along with some simple normalization procedures. Thus, in addition to predicting length, these equations can also predict the mean or expected value of the alignment scores $E(S)$ between random sequences of lengths m and n. Assessing statistical significance then boils down to calculating the probability that an alignment score between two random or unrelated sequences will actually go above $E(S)$. Hence, the expected score or mean extreme score is

$$E(S) = [\log_e\ (Kmn)]/\lambda \qquad ...(13)$$

Another important mathematical result bearing on this question was that the number of matched regions that exceeds the mean score $E(S)$ in Equation 13 could be predicted by the Poisson distribution where the mean x of the Poisson distribution is given by $E(S)$. The Poisson distribution applies when the probability of success in a single trial is small, but the number of trials is large (as in comparing many pairs of random sequences or a test sequence to many scrambled versions of a second sequence) so that some trials end in success but others do not. Some alignments do not reach the expected score, but others will reach or even exceed that score. The Poisson distribution gives the probability P_n of the number of successes, i.e., 0, 1,2,3... when the average number is x and is given by the formula $P_n = e^{-x}/n!$. The probability that no score from many test alignments will exceed

x is therefore approximated by ($P_0 = e^{-x}$). The probability that at least one score exceeds x is $1 - P_0$ and is given by $P(S > x) = 1 - e^{-x}$, so that

$$P(S < x) \simeq \exp(-E(S)) \simeq \exp(-Kmne^{-\lambda x}) \quad ...(14)$$

$$P(S > x) = 1 - \exp(-Kmne^{-\lambda x}) \quad ...(15)$$

Equation 15 estimates the probability of a score greater than x between two random sequences and is identical to the extreme value distribution described below. The Poisson approximation provides a very convenient way to estimate K and λ from alignment scores between many random or unrelated sequences by using the fraction of alignments that have a score less than value x.

Alignments with Gaps

It was predicted on mathematical grounds and shown experimentally that a similar type of analysis holds for sequence alignments that include gaps. Thus, when Smith et al. (1985) optimally aligned a large number of unrelated vertebrate and viral DNA sequences of different lengths (n and m) and their complements to each other, using a dynamic programming local alignment method that allowed for a score of +1 for matches, -0.9 for mismatches, and -2 for a single gap penalty (longer gaps were not considered in order to simplify the analysis), a plot of the similarity score (S) versus the $\log_1/p(nm)$ produced a straight line with approximately constant variance. This result is as expected in the above model except that with the inclusion of gaps, the slope was increased and was of the form

$$S_{mean} = 2.55\ (\log_1/p(mn)) - 8.55 \quad ...(16)$$

with constant standard deviation $\sigma = 1.78$. This result was then used to calculate how many standard deviations were between the predicted mean and variance of the local alignment scores for unrelated sequences and the scores for test pairs of sequences. It the actual alignment score exceeded the predicted S_{mean} by several standard deviations, then the alignment score should be significant. For example, the expected score between two unrelated sequences of lengths 2948 and 431, average $p = 0.279$, was $S_{mean} = 2.55 \times \log_{1/0.279}(2948 \times 431) - 8.99 = 2.55 \times (\log_e(2948 \times 431)/\log_e(1/0.279)) - 8.99 = 2.55 \times 14.1/12.8 - 8.99 = 28.1 - 8.99 = 19.1$. The actual optimal alignment score between the two real sequences of these lengths was 37.20, which exceeds the alignment score expected for random sequences by $(37.20 - 19.1)/1.78 = 10.2\sigma$. Is this number of standard deviations significant? Smith et al. (1985) and Waterman (1989) suggested the use of a conservative statistic known as Chebyshev's inequality, which is valid for many probability distributions: The probability that a random variable exceeds its mean is less than or equal to the square of 1 over the number of standard deviations from the mean. In this example where the actual score is 10 standard deviations above the mean, the probability is $(1/10)^2 = 0.01$.

Waterman (1989) has noted that for low mismatch and gap penalties, e.g., +1 for matches, –0.5 for mismatches, and –0.5 for a single gap penalty, the predicted alignment scores between random sequences as estimated above are not accurate because the score will increase linearly with sequence length instead of with the logarithm of the length. The linear relationship arises when the alignment is more global in nature, and the logarithmic relationship when it is local. Waterman (1989) has fitted alignment scores from a large number of randomly generated DNA sequences of varying lengths to either the predicted $\log(n)$ or n linear relationships expected for low- and high-valued mismatch and gap penalties. The results provide the mean and standard deviation of an alignment score for several scoring schemes, assuming a constant gap penalty.

With further mathematical analysis, it became apparent that the expected scores between alignment of random and unrelated sequences follow a distribution called the Gumbel extreme value distribution. This type of distribution is typical of values that are the highest or best score of a variable, such as the number of heads only expected in a coin toss discussed previously. Subsequently, *S.* Karlin and *S.* Altschul (1990, 1993) further developed the use of this distribution for evaluating the significance of ungapped segments in comparisons between a test sequence and a sequence database using the BLAST program. The method is also used for evaluating the statistical features of repeats and amino acid patterns and clusters in the same sequence. The program SAPS developed by S. Karlin and colleagues at Stanford University and available at http://ulrec3.unil.ch/software/softwarc.html provides this type of analysis. The extreme value distribution is now widely used for evaluating the significance of the score of local alignments of DNA and protein sequence alignments, especially in the context of database similarity searches.

Gumbel Extreme Value Distribution

When two sequences have been aligned optimally, the significance of a local alignment score can be tested on the basis of the distribution of scores expected by aligning two random sequences of the same length and composition as the two test sequences. These random sequence alignment scores follow a distribution called the extreme value distribution, which is somewhat like a normal distribution with a positively skewed tail in the higher score range. When a set of values of a variable are obtained in an experiment, biologists are used to calculating the mean and standard deviation of the entire set assuming that the distribution of values will follow the normal distribution. For sequence alignments, this procedure would be like obtaining many different alignments, both good and bad, and averaging all of the scores. However, biologically interesting alignments are those that give the highest possible scores, and lower scores are not of interest. The experiment, then, is one of obtaining a set of values, and then of using only the highest value and discarding the rest. The focus changes from the statistical approach of wanting to know the average of scores of random sequences, to one of knowing how high a value will be obtained next time another set of alignment scores of random sequences is obtained.

The distribution of alignment scores between random sequences follows the extreme value distribution, not the normal distribution. After many alignments, a probability distribution of highest values will be obtained. The goal is to evaluate the probability that a score between random or unrelated sequences will reach the score found between two real sequences of interest. If that probability is very low, the alignment score between the real sequences is significant and the sequence similarity score is significant.

The probability distribution of highest values in an experiment, the extreme value distribution, is compared to the normal probability distribution in Figure 4.7. The equations giving the respective y coordinate values in these distributions, Y_{ev} and Y_n, are

$$Y_{ev} = \exp\,[-\,x - e^{-x}] \text{ for the extreme value distribution } \quad ...(17)$$

$$Y_n = 1 / \sqrt{(2\pi)}\,\exp[(-x^2)/2] \text{ for the normal distribution } \quad ...(18)$$

The area under both curves is 1. The normal curve is symmetrical about the expectation value or mean at $x = 0$, such that the area under the curve below the mean (0.5) is the same as that above the mean (0.5) and the variance σ^2 is 1. The probability of a particular value of x for the normal distribution is obtained by calculating the area under curve B, usually between $-x$ and $+x$. For $x = 2$, often used as an indication of a significant deviation from the mean, the area between -2 and $+2$ is 0.9544. For the extreme value distribution,

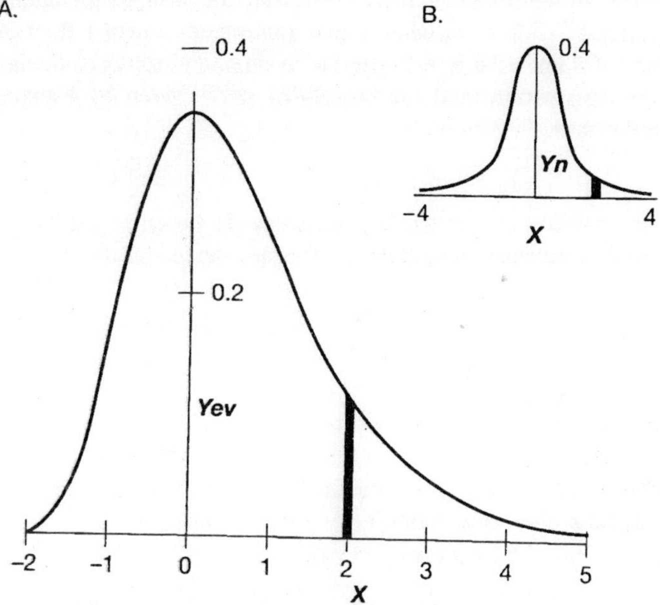

Fig. 4.7. Probability values for the extreme value distribution (A) and the normal distribution (B). The area under each curve is 1.

the expectation value or mean of x is the value of the Euler-Mascheroni constant, 0.57722 ... and the variance of x, σ^2, is the value of $\pi^2/6 = 1.6449$. The probability that score S will be less than value x, $P(S < x)$, is obtained by calculating the area under curve A from $-\infty$ to x, by integration of Equation 17 giving

$$P(S < x) = \exp[-e^{-x}] \qquad \text{...(19)}$$

and the probability of $S \geq x$ is 1 minus this probability

$$P(S \geq x) = 1 - \exp[-e^{-x}] \qquad \text{...(20)}$$

For the extreme value distribution, the area below $x = 0$, which represents the peak or mode of the distribution, is $1/e$ or 0.368 of the total area of 1, and the area above the mean is $1 - 0.368 = 0.632$. At a value of $x = 2$, $Yev = 0.118$ and $P(S < 2) = \exp[-e^{-2}] = 0.873$. Thus, just over 0.87 of the area under the curve is found below $x = 2$. An area of 0.95 is not reached until $x = 3$. The difference between the two distributions becomes even greater for larger values of x. As a result, for a variable whose distribution comes from extreme values, such as random sequence alignment scores, the score must be greater than expected from a normal distribution in order to achieve the same level of significance.

The above equations are modified for use with scores obtained in an analysis. For a variable x that follows the normal distribution, values of x are used to estimate the mean m and standard deviation σ of the distribution, and the probability curve given by Equation 18 then becomes

$$Y_n = 1 / (\sigma\sqrt{(2\pi)}\exp[-(x-m)^2 / 2\sigma^2] \qquad \ldots(21)$$

The probability of a particular value of x can be estimated by using m and σ to estimate the number of standard deviations from the mean, Z, where $Z = (x - m)/\sigma$. Similarly, Equations 17 and 20 can be modified to accommodate the extreme values such as sequence alignment scores

$$P\ (S \geq x) = 1 - \exp\ [-e^{-\lambda(x\ -\ u)}] \qquad \ldots(22)$$

where u is the mode, highest point, or characteristic value of the distribution) and λ is the decay or scale parameter. As is apparent in Equation 22, λ converts the experimentally measured values into standard values of x after subtraction of the mode from each score.

It is quite straightforward to calculate u and λ, and several methods using alignment scores are discussed on the book Web site. There is an important relationship between u and λ, and the mean and standard deviation of a set of extreme values. The mean and standard deviation do not only apply to the normal distribution, but in fact are mathematically defined for any probability distribution. The mean of any set of values of a variable may always be calculated as the sum of the values divided by their number. The mean m or expected value of a variable x, $E\ (x)$, is defined as the first moment of the values of the variable around the mean. From this definition, the mean is that number from which the sum of deviations to all values is zero. The standard deviation σ^2 is the second moment of the values about the mean and is the sum of the squares of the devations from the mean divided by the number of observations less one $(n - 1)$. The mean $\underline{x}$ and standard deviation σ of a set of extreme values can be calculated in the same way, and then u and λ can be calculated using the following equations derived by mathematical evaluation of the first and second moments of the extreme value distribution.

$$\lambda = \pi / (\sigma\sqrt{6}) = 1.2825 / \sigma \qquad \ldots(23)$$

$$u = \underline{x} - \gamma / \lambda = \underline{x} - 0.4500\sigma \qquad \ldots(24)$$

where γ was already introduced. Equation 23 is derived from the ratio of the standard deviations σ^2 of the two distributions or 1 to $\pi^2/6$.

Equation 24 is derived from the observation that the mode or the EV distribution has the value of γ less than the mean. However, the value of γ must be scaled by the ratio of the standard deviations. Hence γ/λ is subtracted from the mean. This method of calculating u and λ from means and standard deviations is called the method of moments.

As with the normal distribution, z scores may be calculated for each extreme value x, where $z = (x - m)/\sigma$ is the number of standard deviations from the mean m to each score. z scores are used by the FASTA, version 3, programs distributed by W. Pearson (1998). Equation 22 may be written in a form that directly uses z scores to evaluate the probability that a particular score Z exceeds a value z,

$$P\,(Z > z) = 1 - \exp\,(-e^{-1.2825\,z\,-0.5772}) \quad ...(25)$$

For sequence analysis, u and λ depend on the length and composition of the sequences being compared, and also on the particular scoring system being used. They can be calculated directly or estimated by making many alignments of random sequences or shuffled natural sequences, using a scoring system that gives local alignments. The parameters will change when a different scoring system is used. Examples of programs that calculate these values are given below.

For alignments that do not include any gaps, u and λ may be calculated from the scoring matrix. The scaling factor λ is calculated as the value of x, which satisfies the condition

$$\Sigma\, p_i\, p_j\, e^{s_{ij}x} = 1 \quad ...(26)$$

where p_i and p_j are the respective fractional representations of residues i and j in the sequences, and s_{ij} is the score for a match being i and j, taken from a log odds scoring matrix, u, the characteristic value of the distribution, is given by

$$u = (\ln Kmn)/\lambda \quad ...(27)$$

where m and n are the sequence lengths and K is a constant that can also be calculated from the values of p_i and s_{ij}. Note that this value originates from the coin toss analysis that gave rise to Equation 14. Combining Equations 25 and 27 eliminates u and gives the following relationship

$$\begin{aligned} P\,(S \geq x) &= 1 - \exp\,[-e^{-\lambda(x\,-u)}] \\ &= 1 - \exp\,[-e^{-\lambda(x\,-\,(\ln Kmn)/\lambda)}] \\ &= 1 - \exp\,[-e^{-\lambda x\,+\,\ln Kmn}] \quad ...(28) \\ &= 1 - \exp\,[-\,kmn\;e^{-\lambda x}] \quad ...(29) \end{aligned}$$

To facilitate calculations, a sequence alignment score S may also he normalized to produce a score S'. The effect of normalization is to change the score distribution with $u = 0$ and $\lambda = 1$. From Equation 28, S′ is calculated by

$$S' = \lambda S - \ln Kmn \quad \text{...(30)}$$

The probability of $P(S' > x)$ is then given by Equation 20 with S = S′

$$P(S' \geq x) = 1 - \exp[-e^{-x}] \quad \text{...(31)}$$

The probability of a particular normalized score may then be readily calculated. This capability depends on a determination of the λ and K to calculate the normalized scores S' by Equation 30.

The probability function $P(S' \geq x)$ decays exponentially in x as x increases and $P(S' \geq x) = 1 - \exp[-e^{-x}] \to e^{-x}$. Consequently, an important approximation for Equations 29 and 31 for the significant part of the extreme value distribution where $x > 2$ is shown in Equations 32 and 33. Note that the replacement equations are single and not double exponentials.

$$P(S \geq x) \simeq Kmn\, e^{-\lambda x} \quad \text{...(32)}$$

$$P(S' \geq x) \simeq e^{-x} \quad \text{...(33)}$$

Table 4.6. Approximation of $P(S' \geq x)$ by e^{-x}

x	$1-\exp[-e^{-x}]$	e^{-x}
0	0.63	1
1	0.308	0.368
2	0.127	0.135
3	0.0486	0.0498
4	0.0181	0.0183

A comparison of probability calculation using this approximation instead of that given in Equation 31 is shown in Table 4.6. For $x > 2$, the estimates differ by less than 2%. The estimate given in Equation 32 also provides a quicker method for estimating the significance of an alignment score.

A Quick Determination of the Significance of an Alignment Score

Scoring matrices are most useful for statistical work if they are scaled in logarithms to the base 2 called bits. Scaling the matrices in this fashion does not alter their ability to score sequence similarities, and thereby to distinguish good matches from poor ones, but does allow a simple estimation of the significance of an alignment. The actual alignment may then be calculated by summing the matrix values for each of the aligned pairs, using matrix values in bit units. If the actual alignment score in bits is greater than expected for alignment of random sequences, the alignment is significant.

For a typical amino acid scoring matrix and protein sequence, K = 0.1 and λ depends on the values of the scoring matrix. If the log odds matrix is in units of bits as described above, then $\lambda = \log_e 2 = 0.693$, and the following simplified form of Equation 32 may be derived by taking logarithms to the base 2 and setting *p* as the probability of the scores of random or unrelated alignments reaching a score of *S* or greater

$$\begin{aligned} \log_2 p &= \log_2 (Kmn\ e^{-\lambda S}) \\ &= \log_2 (Kmn) + \log_2(e^{-\lambda S}) \\ &= \log_2 (Kmn) + (\log_e(e^{-\lambda S}))/\log_e 2 \\ &= \log_2 (Kmn) - \lambda S/\log_e 2 \\ &= \log_2 (Kmn) - S \end{aligned} \qquad \ldots(34)$$

then *S*, the score corresponding to probability *P*, may be obtained by rearranging terms of Equation 34 as follows

$$\begin{aligned} S &= \log_2 (Kmn) - \log_2 P \\ &= \log_2 (K/P) + \log_2(nm) \end{aligned} \qquad \ldots(35)$$

Since for most scoring matrices $K \simeq 0.1$ and choosing $P = 0.05$, the first term is 1, and the second term in Equation 35 becomes the most important one for calculating the score, thus giving

$$S \simeq \log_2(nm) \qquad \ldots(36)$$

Importance of the Type of Scoring Matrix for Statistical Analyses

Using a log odds matrix in bit units simplifies estimation of the significance of an alignment. The Dayhoff PAM matrices, the BLOSUM matrices, and the nucleic acid PAM scoring matrices are examples of this type. Such matrices are also useful for finding local alignments because the matrix includes both positive and negative values. Another important feature of the log odds form of the scoring matrix is that this design is optimal for assessing statistical significance of alignment scores. A set of matrices, each designed to detect similarity between sequences at a particular level, is best for this purpose. Use of a matrix that is designed for aligning sequences that have a particular level of similarity (or evolutionary distance) assures the highest-scoring alignment and therefore the very best estimate of significance. Thus, lower-numbered PAM matrices are most suitable for aligning sequences that are more similar. In the above example, the Dayhoff PAM250 matrix designed for sequences that are 20% similar was used to align sequences that are approximately 20% identical and 50% similar (identities plus common replacements in the alignment). Using a lower

PAM120 matrix produces a slightly higher score for this alignment, and thus increases the significance of the alignment score.

Another important parameter of the scoring matrix for statistical purposes is the expected value of the average amino acid pair, calculated as shown in Equation 37. This value should be negative if alignment scores for the matrix are to be used for statistical tests, as performed in the above example. Otherwise, in any aligned pair of sequences the scores will increase with length faster than the logarithm of the length. Not all scoring matrices will meet this requirement. To calculate the expected score (E), the score for each amino acid pair (s_{ij}) is multiplied by the fractional occurrences of each amino acid (p_i and p_j). This weighted score is then summed over all of the amino acid pairs. The expected values of the log odds matrices such as the Dayhoff PAM, BLOSUM, JTT, JO93, PET91, and Gonnet92 matrices all meet this statistical requirement.

$$E = \sum_{i=1}^{20} \sum_{j=1}^{i} p_i q_i \, s_{ij} \qquad \text{...(37)}$$

For example, for the PAM120 matrix in one-half bits $E = -1.64$ and for PAM160 in one-half bits, $E = -1.14$. Thus, scores obtained with these matrices may be used in the above statistical analysis. Ungapped alignment scores obtained using the BLOSUM62 matrix may also be subject to a significance test, as described above for the PAM matrices. The test is valid because the expect score for a random pair of amino acids is negative (E = –0.52). Because the matrix is in half-bit units, the alignment is significant when a score exceeds $16/0.52 \approx 32$ half-bits.

To assist in keeping track of information, scoring matrices have appeared in a new format suitable for use by many types of programs. The matrix includes: (1) the scale of the matrix and the value of the statistical parameter λ; (2) E, the expect score of the average amino acid pair in the matrix, which if negative assures that local alignments will be emphasized (Eq. 37); (3) H, the information content or entropy of the matrix (Eq. 3) giving the ability of the matrix to discriminate related from unrelated sequence alignments, not shown here; and (4) suitable gap penalties. The BLOSUM matrices are also available in this same format.

Significance of Gapped, Local Alignments

When random sequences of varying lengths are optimally aligned with the Smith-Waterman dynamic programming algorithm using an appropriate scoring matrix and gap penalties, the distribution of scores also matches the extreme value distribution. Similarly, in optimally aligning a given sequence to a database of sequence; and after removing the high scores of the closely related sequences, the scores of the unrelated sequences also follow this distribution. In these and other cases, optimal scores are found to increase linearly with log (n), where n is the sequence length. Equation 33 predicts that the optimal alignment score (x) expected between two random or unrelated sequences should be proportional to the logarithm of the product of the sequence lengths, $x \simeq \log_2(nm)$. If the sequence lengths are approximately equal, $n \simeq m$, then x should be proportional to $\log_2(n^2) = 2 \log_2(n)$, and the predicted score should also increase linearly with $\log(n).\log_2(n)$ is equivalent to $\log(n)$ because, to change the base of a logarithm, one merely multiplies by a constant. In comparing one sequence of length m to a sequence database of length n, m is a constant and the predicted score should increase linearly as $\log(n)$. This $\log(n)$ relationship has been found in several studies of the distribution of optimal local alignment scores that have included gap penalties. Thus, the same statistical methods described above for assessing the significance of ungapped alignment scores may also be used for gapped alignment scores. Methods for calculating the parameters K and λ for a given combination of scoring matrix methods and gap penalties are described on the book Web site.

Methods for Calculating the Parameters of the Extreme Value Distribution

In the analysis by Altschul and Gish (1996), 10,000 random amino acid sequences of variable lengths were aligned using the Smith-Waterman method and a combination of the scoring matrix and a reasonable set of gap penalties for the matrix. The scores found by this method followed the same extreme value distribution predicted by the underlying statistical theory. Values of K and λ were then estimated for each combination by fitting the data to the predicted extreme value distribution.

Altschul and Gish (1996) have cautioned users of these statistical parameters. First, the parameters were generated by alignment of random sequences that were produced assuming a particular amino acid distribution, which may be a poor model for some proteins. Second, the accuracy of λ and K cannot be estimated easily. Finally, for gap costs that give values of $H < 0.15$, the optimal alignment length is a significant traction of the sequence lengths and produces a source of error called the edge effect. The effect occurs when the expected length of an alignment is a significant fraction of the sequence length, and, as discussed earlier, alignments between sequences that overlap at their ends cannot be completed. The expected length is then subtracted from the sequence length before λ is estimated. If no such correction is done, λ may be overestimated.

These values for gap penalties should also not be construed to represent the best choice for a given pair of sequences or the only choices, simply because the statistical parameters are available. The process of choosing a gap penalty remains a matter of reasoned choice. In trying the effects of varying the gap penalty, it is important to recognize that as the gap penalty is lowered, the alignments produced will have more gaps and will eventually change from a local to a global type of alignment, even though a local alignment program is being used. In contrast, higher H values are generated by a very large gap penalty and produce alignments with no gaps, thus suggesting an increased ability to discriminate between related and unrelated sequences. In this respect, Altschul and Gish (1996) note that beyond a certain point increasing the gap extension penalty does not change the parameters, indicating that most gaps in their simulations are probably of length 1. However, reducing the gap penalty can also allow an alignment to be extended and create a higher scoring alignment. Eventually, however, the optimal local alignment score between unrelated sequences will lose the log length relationship with sequence length and become a linear function. At this point, gap penalties are no longer useful for obtaining local alignments and the above statistical relationships are no longer valid.

The higher the H value, the better the matrix can distinguish related from unrelated sequences. The lower the value of H, the longer the expected alignment. These conditions may be better if a longer alignment region is required, such as testing a structural or functional model of a sequence by producing an alignment. Conversely, scoring parameters giving higher values of H should produce shorter, more compact alignments. If $H < 0.15$, the alignments may be very long. In this case, the sequences have a shorter effective length since alignments starting near the ends of the sequences may not be completed. This edge effect can lead to an overestimation of λ but was corrected for in the above table.

Unfortunately, the above method for calculating the significance of an alignment score may not be used to test the significance of a global alignment score. The theory does not apply when these same substitution matrices are used for global alignments. Transformation of these matrices by adding a fixed constant value to each entry or by multiplying each value by a constant has no effect on the relative scores of a series of global alignments. Hence, there is no theoretical basis for a statistical analysis of such scores as there is for local alignments.

Two programs are commonly used for database similarity searches: FASTA and BLAST. These programs both calculate the statistical significance of the higher scores found with similar sequences, but the types of analyses used to determine the statistical significance of these scores are somewhat different. BLAST uses the value of K and λ found by aligning random sequences and Equation 29, where n and m are shortened to compensate for inability of ends to align. FASTA calculates the statistical significance using the distribution of scores with unrelated sequences found during the database search. In effect, the mean and standard deviation of the low scores found in a given length range are calculated. These scores represent the expected range of scores of unrelated sequences for that sequence length (recall that the local alignment scores increase as the logarithm of the sequence length). The number of standard deviations to the high scores of related sequences in the same length range (z score) is then determined. The significance of this z score is then calculated according to the extreme value distribution expected of the z scores, given in Equation 25. Pearson (1996) showed that these two methods are equally useful in database similarity searches for detecting sequences more distantly related to the input query sequence.

Pearson (1996) has also determined the influence of scoring matrices and gap penalties on alignment scores of moderately related and distantly related protein sequences in the same family. For two examples of moderately related sequences, the choice of scoring matrix and gap penalties (gap opening penalty followed by penalty for each additional gap position) did not matter, i.e., BLOSUM50 –12/–2, BLOSUM62 -8/–2, Gonnet93 –10/–2, and PAM250 –12, –2 all produced statistically significant scores. The scores of distantly related proteins in the same family depended more on the choice of scoring matrix and gap penalty, and some scores were significant and others were not. Pearson recommends using caution in evaluating alignment scores using only one particular combination of scoring matrix and gap penalties. He also suggests that using a larger gap penalty, e.g., –14, –2 with BLOSUM50, can increase the selectivity of a database search for similarity (fewer sequences known to he unrelated will receive a significant alignment score).

A difficulty encountered by FASTA in calculating statistical parameters during a database search is that of distinguishing unrelated from related sequences, because only scores of unrelated sequences must be used. As score and sequence length information is accumulated during the search, the scores will include high, intermediate, and sometimes low scores of sequences that are related to the query sequence, as well as low scores and sometimes intermediate and even high scores of unrelated sequences. As an example, a high score with an unrelated database sequence can occur because the database sequence has a region of low complexity, such as a high proportion of one amino acid. Regardless of the reason, these high scores must be primed from the search if accurate statistical estimates are to be made. Pearson (1998) has devised several such pruning schemes, and then determined the influence of the scheme on the success of a database search at demonstrating statistically significant alignment scores among members of the same protein family or superfamily. However, no particular scheme proved to be better than another.

The above method does not necessarily ensure that the choice of scoring matrix and gap penalties provides a realistic set of local alignment scores. In the comparable situation of matching a test sequence to a database of sequences, the scores also follow the extreme value distribution. For this situation, Mott (1992) has explained that for local alignments the end point of the alignment should on the average be half-way along the query sequence, and for global alignments, the end point should be beyond that half-way point. Pearson (1996) has pointed out that the presence of known, unrelated sequences in the upper part of the curve where $E > 1$ can be an indication of an inappropriate scoring system.

Statistical Significance of Individual Alignment Scores between Sequences and the Significance of Scores Found in a Database Search are Calculated Differently

In performing a database search between a query sequence and a sequence database, a new comparison is made for each sequence in the database. Alignment scores between unrelated sequences are employed by FASTA to calculate the parameters of the extreme value distribution. The probability that scores between unrelated sequences could reach as high as those found for matched sequences can then be calculated. Similarly, in the database similarity search program BLAST, estimates of the statistical parameters are calculated based on the scoring matrix and sequence composition. The parameters are then used to calculate the probability of finding conserved patterns by chance alignment of unrelated sequences. When performing such database searches, many trials are made in order to find the most strongly matching sequences.

As more and more comparisons between unrelated sequences are made, the chance that one of the alignment scores will be the highest one yet found increases. The probability of finding a match therefore has to be higher than the value calculated for a score of one sequence pair. The length of the query sequence is about the same as it would be in a normal sequence alignment, but the effective database sequence is very large and represents many different sequences, each one a different test alignment. Theory shows that the Poisson distribution should apply, as it did above for estimating the parameters of the extreme value distribution from many alignments between random sequences.

The probability of observing, in a database of D sequences, no alignments with scores higher than the mean of the highest possible local alignment scores s is given by e^{-Ds}, and that of observing at least one score s is $P \simeq 1 - e^{-Ds}$. For the range of values of P that are of interest, i.e., $P < 0.1$, $P = Ds$. If two sequences are aligned by PRSS as given in the above example, and the significance of the alignment is calculated, two scores must be considered. The probability of the score may first be calculated using the estimates of λ and K. Thus, in the phage repressor alignment, $P(s > 401) = 3.7. \times 10^{-27}$. However, to estimate the EV parameters, 1000 shuffled sequences were compared, and the probability that one of those sequences would score as high as 401 is given by Ds, or $1000 \times 3.7 \times 10^{-27} = 3.7 \times 10^{-24}$. These numbers are also shown in the statistical estimates computed by PRSS. Finally, if the score had arisen from a database search of 50,000 sequences, the probability of a score of 401 among this many sequence alignments is 5×10^{-19}, still a small number, but 50,000 larger than that for a single comparison. These probability calculations are used for reporting the significance of scores with database sequences by FASTA and BLAST.

Sequence Alignment and Evolutionary Distance Estimation by Bayesian Statistical Methods

A recent development in sequence alignment methods is the use of Bayesian statistical methods to produce alignments between pairs of sequences and to calculate distances between sequences. Before discussing these methods further, we provide some introductory comments about Bayesian probability.

Introduction to Bayesian Statistics

Bayesian statistical methods differ from other types of statistics by the use of conditional probabilities. These probabilities are used to derive the joint probability of two events or conditions. An example of a conditional probability is $P(\mathrm{B}|\mathrm{A})$, meaning the probability of B, given A, whereas $P(\mathrm{B})$ is the probability of B, regardless of the value of A. Suppose that A can have two states, A1 and A2, and that B can also have two states, B1 and B2. These states might, for instance, correspond to two allelic states of two genes. Then, $P(\mathrm{B}) = P(\mathrm{B1}) + P(\mathrm{B2}) = 1$ and $P(\mathrm{A}) = \mathrm{P(A1)} + \mathrm{P(A2)} = 1$. Suppose, further, that the probability $P(\mathrm{B1}) = 0.3$ is known. Hence $P(\mathrm{B2}) = 1 - 0.3 = 0.7$. In our genetic example, each probability might correspond to the frequency of an allele, for which p and q are often used. These probabilities $P(\mathrm{B1})$, etc., can be placed along the right margins of the table as the respective sum of each row or column and are referred to as the marginal probabilities.

Interest is now focused on filling in the missing data in the middle two columns of the table. The probability of A1 and Bl occurring together (the value to be entered in row Bl and column Al) is called the joint probability, $P(\mathrm{B1}$ and $\mathrm{Al})$ (also denoted $P[\mathrm{B1}, \mathrm{A1}])$. The marginal probability $P(\mathrm{A1})$ is also missing. The available information up to this point, called the prior information, is not enough to calculate the joint probabilities. With additional data on the co-occurrence of A1 with B1, etc., these joint probabilities may be derived by Bayes' rule. Suppose that the conditional probabilities $P(\mathrm{A1}|\mathrm{B1}) = 0.8$ and $\mathrm{P(A2|B2)} = 0.70$ are known, the first representing, for example, the proportion of a population with allele B1 that also has allele A1. First, note that $P(\mathrm{A1}|\mathrm{B1}) + P(\mathrm{A2}|\mathrm{B1}) = 1$, and hence that $P(\mathrm{A2}|\mathrm{B1}) = 1.0\text{–}0.8 = 0.2$. Similarly, $P(\mathrm{A1}|\mathrm{B2}) = 1.0 - 0.70 = 0.3$. Then the joint probabilities and other conditional probabilities may be calculated by Bayes' rule, illustrated using the joint probability for A1 and B1 as an example.

$$P(\mathrm{A1 \ and \ B1}) = P(\mathrm{B1})\, P(\mathrm{A1}|\mathrm{B1}) \qquad \text{...(38)}$$

$$P(\mathrm{A1 \ and \ B1}) = \mathrm{P(A1)}\, P(\mathrm{B1}|\mathrm{A1}) \qquad \text{...(39)}$$

Thus, $P(\mathrm{A1 \ and \ B1}) = P(\mathrm{B1}) \times P(\mathrm{A1}|\mathrm{B1}) = 0.3 \times 0.8 = 0.24$, and $P(\mathrm{A2 \ and \ B2}) = P(\mathrm{B2}) \times P(\mathrm{A2}|\mathrm{B2}) = 0.7 \times 0.7 = 0.49$. The other joint probabilities may be calculated by subtraction; e.g., $P(\mathrm{A1}$ and

B2) = P(B1) - P(A1 and B1) = 0.30 - 0.24 = 0.06. To calculate P(A1) and P(A2), the joint probabilities in each column may be added, thereby completing the additions to the table.

However, note that P(A1) may also be calculated in the following manner,

$$P(\text{A1}) = P(\text{A1 and B1}) + P(\text{A1 and B2})$$
$$= P(\text{B1})\ P(\text{A1}|\text{B1}) + P(\text{B2})\ P(\text{A1}|\text{B2}) \quad \text{...(40)}$$

Other conditional probabilities may be calculated from Equations 38 and 39 by rearranging terms and by substituting Equation 40, and the following form of Bayes' rule may be derived,

$$P(B2|\text{A1}) = P(\text{A1 and B2})/P(\text{A1})$$
$$= P(\text{B2})\ P(\text{A1}|\text{B2})/P(\text{A1})$$
$$= \text{P(B2) P(A1}|\text{B2})/[P(\text{B1})\ P(\text{A1}|\text{B1}) + P(\text{B2})\ P(\text{A1}|\text{B2})] \text{..(41)}$$

Using Equation 41, P(B2|A1) = 0.7 × 0.30 / [0.3 × 0.80 + 0.7 × 0.3] = 0.467, and also P(B1|A1) = 1.0 - 0.467 = 0.533. Such calculated probabilities are called posterior probabilities or posteriors, as opposed to the prior probabilities or priors initially available. Thus, based on the priors and additional information, application of Bayes' rule allows the calculation of posterior estimates of probabilities not initially available. This procedure of predicting probability relationships among variables may be repeated as more data are collected, with the existing model providing the prior information and the new data providing the information to derive a new model. The initial beliefs concerning a parameter of interest are expressed as a prior distribution of the parameter, the new data provide a likelihood for the parameter, and the normalized product of the prior and likelihood (Eq. 41) forms the posterior distribution.

In the above example, note that the joint probability of A1 and B1 [P(A1 and B1)] is not equal to the product of P(A1) and P(B1); i.e., 0.24 is not equal to 0.3 × 0.45 = 0.135. Such would be the case if the states of A and B were completely independent; i.e., if A and B were statistically independent variables as, for example, in a genetic case of two unlinked genes A and B. In the above example, the state of one variable is influencing the state of the other such that they are not independent of each other, as might be expected for linked genes in the genetic example.

A more general application of Bayes' rule is to consider the influence of several variables on the probability of an outcome. The analysis is essentially the same as that outlined above. To see how the method works with three instead of two values of a variable, think first of an example of three genes, each having three alleles, and of deriving the corresponding conditional probabilities. The resulting joint probabilities will depend on the choice made of the three possible values for each variable. To go even farther, instead of a small number of discrete sets of alternative values of a variable, Bayesian statistical methods may also be used with a large number of values of variables or even with continuous variables.

For sequence analysis by Bayesian methods, a slightly different approach is taken. The variables may include combinations of possible alignments, gap scoring systems, and log odds substitution matrices. The most probable alignments may then be identified. The scoring system used for sequence alignments is quite readily adapted to such an analysis. In an earlier discussion, it was pointed out that a sequence alignment score in bits is the logarithm to the base 2 of the likelihood of obtaining the score in alignments of related sequences divided by the likelihood of obtaining the score in alignments of unrelated sequences. It was also indicated that the highest alignment score should be obtained if the scoring matrix is used that best represents the nucleotide or amino acid substitutions expected between sequences at the same level of evolutionary distance. Bayesian methodology carries this analysis one step farther by examining the probabilities of all possible alignments of the sequences using all possible variations of the input parameters and matrices. These selections are the prior information for the Bayesian statistical analysis and provide various estimates of the alignment that allow us to decide on the most probable alignments. The alignment score for each combination of these variables provides an estimate of the probability of the alignment. By using equations of conditional probability such as Equation 41, posterior information on the probability of alignments, gap scoring system, and substitution matrix can be obtained.

Application of Bayesian Statistics to Sequence Analysis

To use an example from sequence analysis, a local alignment score (s) between two sequences varies with the choice of scoring matrix and a gap scoring system. In the previous sections, an amino acid scoring matrix was chosen on the basis of its performance in identifying related sequences. Gap penalties were then chosen for a particular scoring matrix on the basis of their performance in identifying known sequence relationships and of their keeping a local alignment behaviour by the increase in score between unrelated sequences remaining a logarithmic function of sequence length. The alignment score expressed in bit units was the ratio of the alignment score expected between related sequences to that expected between unrelated sequences, expressed as a logarithm to the base 2. The scores may be converted to an odds ratio (r) using the formula $r = 2^s$. The probability of such a score between unrelated or random sequences can then be calculated using the parameters for the extreme value distribution for that combination of scoring matrix and gap penalty. Finally, the above analysis may provide several different alignments, without providing any information as to which is the most likely. With the application of Bayesian statistics, the approach is different.

The application of Bayesian statistics to this problem allows one to examine the effect of prior information, such as the chosen amino acid substitution matrix, on the probability that two sequences are homologous. The method provides a posterior probability distribution of all alignments taking into account all possible scoring systems. Thus, the most likely alignments and their probabilities may be determined. This method circumvents the need to choose a particular scoring matrix and gap scoring system because a range of available choices can be tested. The approach also provides conditional posterior distributions on the gap number and substitution matrix. Another application of Bayes statistics for sequence analysis is to find the PAM DNA substitution matrix that provides the maximum probability of a given level of mismatches in a sequence alignment, and thus to 'predict the evolutionary distance between the sequences.

Bayesian Evolutionary Distance

Agarwal and States (1996) have applied Bayesian methods to provide the best estimate of the evolutionary distance between two DNA sequences. The examples used are sequences of the same length that have a certain level of mismatches. Consequently, there are no gaps in the alignment between the sequences. Sequences of this type originated from gene duplication events in the yeast and *Caenorhabditis elegans* genomes. When there are multiple mismatches between such repeated sequences, it is difficult to determine the most likely length of the repeats. With the application of Bayesian methods, the most probable repeat length and evolutionary time since the repeat was formed may be derived.

The alignment score in bits between sequences of this type may be calculated from the values for matches and mismatches in the DNA PAM scoring matrices described earlier. Recall that a PAM1 evolutionary distance represents a change of 1 sequence position in 100 and is thought to correspond roughly to an evolutionary distance of 10^7 years. Higher PAMN tables are calculated by multiplying the PAM1 scoring matrix by itself *n* times. This Markovian model of evolution assumes that any sequence position can change with equal probability, and subsequent changes at a site are not influenced by preceding changes at that site. In addition, a changed position can revert to the original nucleotide at that position. The problem is to discover which scoring matrix (PAM50, 100, etc.) gives the most likely alignment score between the sequences. This corresponding evolutionary distance will then represent the time at which the sequence duplication event could have occurred.

An approach described earlier was to evaluate the alignment scores using a series of matrices and then to identify the matrix giving the highest similarity score. For example, if there are 60 mismatches between sequences that are 100 nucleotides long, the PAM50 matrix score of the alignment in bits ($\log_2$) is $40 \times 1.34 - 60 \times 1.04 = -8.8$, but the PAM125 matrix score is much higher, $40 \times 0.65 - 60 \times 0.30 = 8$. When these log odds scores in bits are converted to odds scores, the difference is 0.002 versus 256. Thus, the PAM125 matrix provides a much better estimate of the evolutionary distance between sequences that have diverged to this degree. The Bayesian approach continues this type of analysis to discover the probability of the alignment as a function of each evolutionary distance represented by a different PAM matrix, If *x* is the evolutionary distance represented by the PAMN matrix divided by 100, and *k* is the number of mismatches in a sequence of length *n*, then by Bayes' rule and related formulas discussed above

$$P(x|k) = P(k|x)\,P(x)/P(k)$$
$$= P(k|x)P(x)/\textstyle\sum_x P(k|x) \qquad \ldots(42)$$

$P(x|k)$ is the probability at distance *x* given the sequence with *k* mismatches (and $n - k$ matches), $P(k|x)$ is the odds score for the sequence with *k* mismatches using the log odds scores in the DNA PAM100*x* matrix, and $P(x)$ is the prior probability of distance *x* (usually 1 over the number of matrices, thus making each one equally possible). The denominator is the sum of the odds scores over the range of *x*, which is 0.01 – 4, representing PAM1 to PAM400 (~10 million to 4 billion years). Like the conditional probabilities calculated by Equation 42, this sum represents the area under the probability curve and has the effect of normalizing the probability for each individual scoring

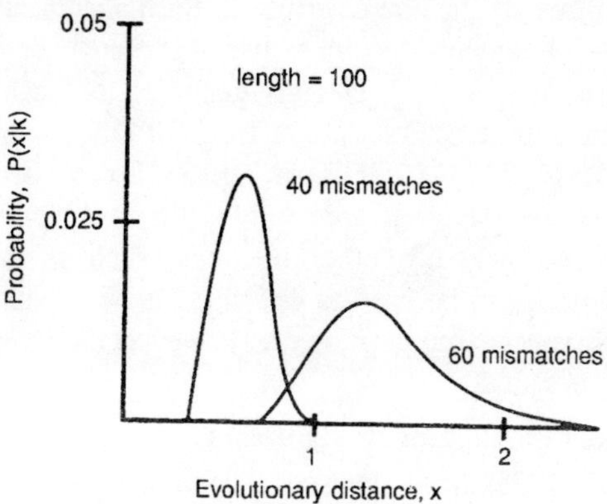

Fig. 4.8. P(x | k) for sequence length n=100 and number of mismatches k=40 or 60.

matrix used. The shape of the probability curve reveals how $P(x|k)$ varies with x.

The probability curves have a single mode or highest score for $k < 3n/4$. Because the curves are not symmetrical about this mode but are skewed toward higher distances, the expected value or mean of the distribution and its standard deviation are the best indication of evolutionary distance. For a sequence 100 nucleotides long with 40 mismatches, the expected value of x is 0.60 with $s = 0.11$, representing a distance of ~ 600 million years. These estimates are different from the earlier method that was described of finding the matrix that gives the highest alignment score, which would correspond to the mode or highest scoring distance.

One difficulty with making such estimations is that the estimate depends on the assumption that the mutation rate in sequences has been constant with time (the molecular clock hypothesis) and that the rate of mutation of all nucleotides is the same. Such problems may be solved by scoring different portions of a sequence with a different scoring matrix, and then using the above Bayesian methods to calculate the best evolutionary distance. Another difficulty is deciding on the length of sequence that was duplicated. In genomes, the presence of repeats may be revealed by long regions of matched sequence positions dispersed among regions of sequence positions that do not match. However, as the frequency of mismatches is increased, it becomes difficult to determine the extent of the repeated region. The application of the above Bayesian analysis allows a determination of the probability distributions as a function of both length of the repeated region and evolutionary distance. A length and distance that gives the highest overall probability may then be determined. Such alignments are initially found using an alignment algorithm and a particular scoring matrix. Analysis of the yeast and *C. elegans* genomes for such repeats has underscored the importance of using a range of DNA scoring matrices such as PAM1 to PAM120 if most repeats are to be found. One disadvantage of the Bayesian approach is that a specific mutational model is required, whereas other methods, such as the maximum likelihood approach can be used to estimate the best mutational model as well as the distance. Computationally, however, the Bayesian method is much more practical.

Bayesian Sequence Alignment Algorithms

Zhu et al. (1998) have devised a computer program called the Bayes block aligner which in effect slides two sequences along each other to find the highest scoring ungapped regions or blocks. These blocks are then joined in various combinations to produce alignments. There is no need for gap penalties because only the aligned sequence positions in blocks are scored. Instead of using a given substitution matrix and gap scoring system to find the highest scoring alignment, a Bayesian statistical approach is used. Given a range of substitution matrices and number of blocks expected in an alignment as the prior information, the method provides posterior probability distributions of alignments. A graphical interface for X windows in a UNIX environment and a nongraphical interface for PCs running Windows are available. The method may be used for both protein and DNA sequences. An alignment block between two sequences is defined as a run of one or more identical characters in the sequence alignment that can include intervening mismatches but no gaps, as shown in the following example. Only the aligned blocks are identified and scored; regions of unaligned sequence and gaps between these blocks are not scored. The probability of a given alignment is given by the product of the probabilities of the individual alignment scores in the blocks, as indicated in the following example. The Bayes block aligner scores every possible combination of blocks to find the best scoring alignment.

Another type of analysis that can be performed with the Bayes block aligner is to examine the probability of the alignments. The procedure is entirely different from other methods of sequence alignment such as dynamic programming. On the one hand, with dynamic programming methodology, a single best alignment is found for a given scoring matrix and gap penalty, and the odds for finding as good a score between random sequences of the same length and complexity is

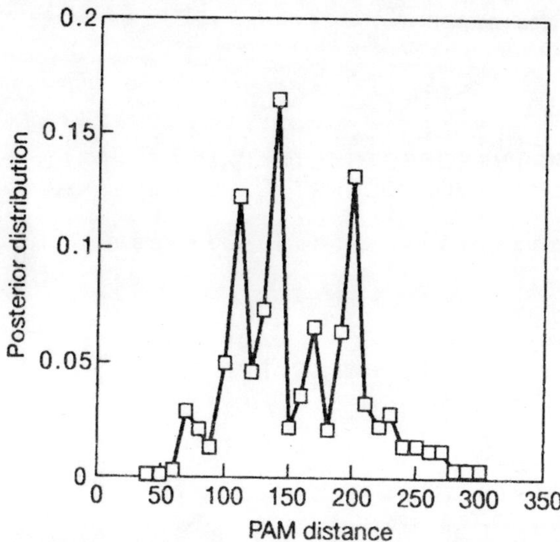

Fig. 4.9. Posterior probability distribution of Dayhoff PAM scoring matrices for alignment of 1GKY and 2AK3-A.

determined. On the other hand, with Bayesian alignment methods, all possible alignments are considered for a reasonable number of blocks and a set of substitution matrices. Rather than a probability of a single alignment, the probabilities of many alignments are provided. Many possible alignments may be examined and compared, and the frequency of certain residues in the sequences in these alignments may be determined.

For 1GKY and 2AK3-A, no highly probable single optimal or near-optimal alignment is found, suggesting these alignments are not representative of the best possible alignment of these sequences. Experience with the method has suggested that a minimum number of blocks that best represents the expected domain structure is the best approach. An average alignment for a number of blocks of probability greater than 0.9 has been found to give good agreement with predicted structural alignments. Values of k are obtained from the probability distribution for k. Using this approach with the Bayes aligner, the alignments between 1GKY and 2AK3-A have been predicted. Although most of the predicted alignments correspond to expected structural alignments with the active site of the enzyme, alignment II does not so correspond. Such false-negative predictions of structural alignments are the commonest error of Bayesian methods, probably because of relaxed conditions for scoring alignments in the use of unconstrained

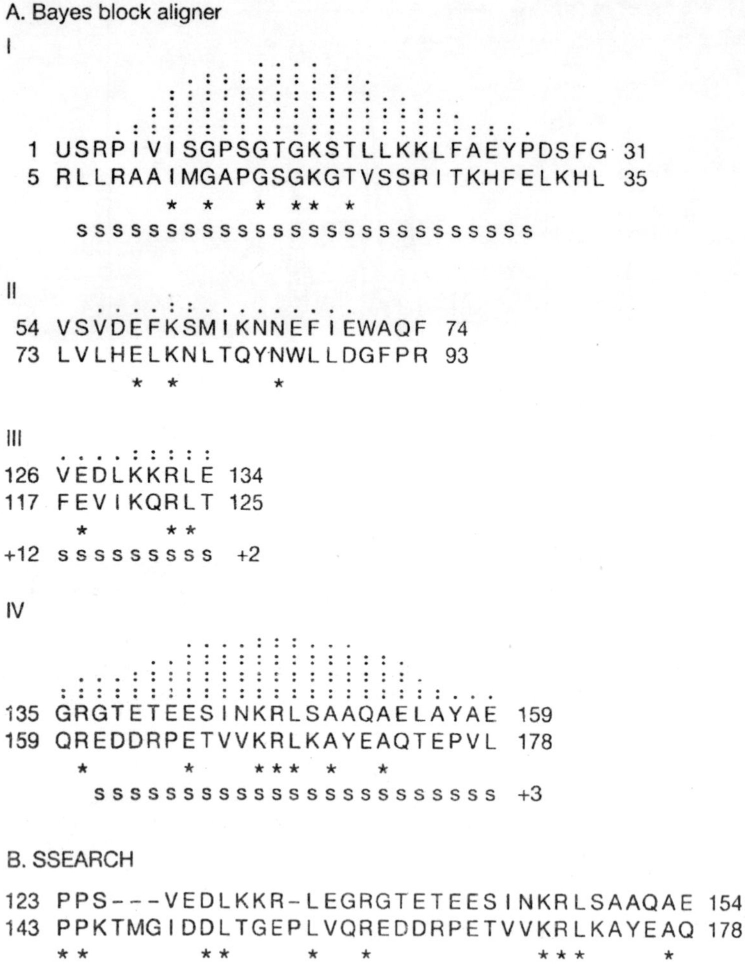

A. Bayes block aligner

```
I

  1 USRPIVISGPSGTGKSTLLKKLFAEYPDSFG 31
  5 RLLRAAIMGAPGSGKGTVSSRITKHFELKHL 35
           *  *   *  **  *
    ssssssssssssssssssssssssssss

II

 54 VSVDEFKSMIKNNEFIEWAQF 74
 73 LVLHELKNLTQYNWLLDGFPR 93
        *  *      *

III

126 VEDLKKRLE 134
117 FEVIKQRLT 125
      *     **
+12 sssssssss +2

IV

135 GRGTETEESINKRLSAAQAELAYAE 159
159 QREDDRPETVVKRLKAYEAQTEPVL 178
      *      *   ***  *   *
       sssssssssssssssssssssss +3
```

B. SSEARCH

```
123 PPS---VEDLKKR-LEGRGTETEESINKRLSAAQAE 154
143 PPKTMGIDDLTGEPLVQREDDRPETVVKRLKAYEAQ 178
    **        **    *  *        ***    *
```

Fig. 4.10. The alignment of 1GKY and 2AK3-A obtained with the Bayes aligner (A) and by SSEARCH (B), a dynamic programming method that provides local alignments.

prior information. For these proteins, which share little sequence identity, the Bayes aligner correctly predicts many, but not all, features of the structural alignment, and does so better than a dynamic programming method that provides local alignments. In other cases, the Bayes aligner may not perform as well as dynamic programming. The prudent choice is to use the Bayes aligner as one of several computer tools for aligning sequences.

5

USING PERL TO FACILITATE BIOLOGICAL ANALYSIS

Consider a situation in which an investigator is studying genes that affect neuronal signaling in *C. elegans*, with a primary interest in identifying those with gene products that may be secreted. A Web site that reports the results of a large systematic study of predicted *C. elegans* genes using the RNA-induced inhibition of gene expression (RNAi) technique is available, and the investigator can download a summary file based on a few thousand experiments. The latest release of WormPep, which reports peptide sequences of over 19,000 known and predicted worm genes, is also available. What is now needed is the ability to search the RNAi results file for those genes that affect worm movement in some way (e.g., a common phenotype for genes affecting neuronal activity), and extract the sequence of those genes from WormPep. The ultimate plan is to submit these sequences of interest to SignalP, an E-mail-based signal peptide cleavage site predictor based at the Technical University of Denmark.

How would one accomplish this? One way is to do the job by hand. First, one would need to read the RNAi summary file into a word processing program, cull it for experiments that affected locomotion in some way, and then assemble a list of all of the genes that produce relevant phenotypes. Next, one would open WormPep, search for the corresponding sequence for each of these genes, and cut-and-paste these sequences into another file. The last step would be to reformat the entries into the format required by the SignalP server, pasting all of the entries into an E-mail message.

It should be immediately apparent that there are some problems with this approach. If there are more than just a few genes of interest to analyze, the job quickly becomes rather tedious, if not overly time consuming. Worse yet, the next time that new RNAi results are released, the whole process will need to be repeated, determining which RNAi entries are new. The step involving loading WormPep into a word processor might not even be tenable due to the sheer size of the database, well over 10 megabases.

This is the type of problem that a Perl script can help with. The Perl programming language excels at slicing, dicing, and integrating data files and is the language of choice for the many bioinformatics researchers. This chapter will provide a gentle introduction to Perl, with examples designed to illustrate the usefulness of learning this language.

GETTING STARTED

During the course of this chapter, a solution will be developed for the data integration problem introduced in the first paragraph of this chapter. Before this problem is attacked, however, some very simple scripts that illustrate the basics of Perl programming will be discussed. In considering these examples, the reader is strongly encouraged to follow along by typing the examples into a Perl interpreter to get a better sense for what these short scripts actually do. Modifying and experimenting with the scripts for individual use is also encouraged.

Perl interpreters are available for the Macintosh, Windows, and UNIX operating systems and more often than not is made available as freeware. A number of download sites are listed at the end of this chapter. Because Perl program files are usually compressed to speed downloads, Macintosh users will need to use a utility like UnStuffIt to uncompress the package, whereas Windows users will need WinZip or PKZip. Perl should come as part of the standard software installation on most UNIX-based machines. Typing `perl -v` at the UNIX prompt will indicate whether the program is indeed installed and, if so, the version number that is available on that machine.

Perl consists of two essential parts: (i) the *interpreter*, called perl on UNIX systems, MacPerl on Macintoshes, and perl.exe on Windows machines: and (ii) *scripts*, text files that are written by the user describing a discrete set of steps to be performed by the interpreter. The scripts are actually computer programs, and the words script and program can be used interchangeably.

The process of writing and running a Perl script is similar on Windows and UNIX systems but slightly different for Macintosh users. The basic steps are described below.

On a Windows or UNIX system, create a text-only file containing the following lines:

```
#!/usr/bin/perl
print "My first Perl script.";
```

Any word processing program may be used, as long as the file is saved as text only. The Windows Notepad program is good for this task, as it saves its files in text-only format by default. Name the file first.pl.

This newly created file contains two lines. The first line is a comment that identifies the file as a Perl script (indicated by the # ! at the beginning of the line). The second line is a *print statement* that tells Perl to print out the text

```
My first Perl script
```

The name of the file ends in the extension .pl, which is a standard naming convention for Perl scripts.

To run this command, open a command-line window (the DOS window on Windows systems, a shell window on UNIX), change to the directory that contains the file, and type the command

```
% perl first.pl
My first Perl script.
```

where % represents the command-line prompts for both the Windows and UNIX systems, and boldface type represents input typed by the user at the keyboard.

What this particular command does is invoke the Perl command interpreter, passing to it the name of the file that should be run. The interpreter dutifully processes the script line-by-line, sees the single print command, and executes it. The output of the script appears in the command window.

If Perl is installed in the standard way on the computer, the `perl` command does not have to be explicitly typed. On UNIX systems, the script file can be marked as being directly executable using the chmod command:

```
% chmod +x first.pl
```

On Windows systems, the file does not have to be explicitly marked as executable because Perl is usually installed in such a way that any file ending in the .pl extension is associated with the Perl interpreter. Simply typing in the name of the program will run it:

```
% first.pl
My first Perl script.
```

Mac-based Perl does not have a command window, so the process of creating and running a Perl script is somewhat different. On Macintoshes, the MacPerl application is launched by double clicking on its icon (a pyramid with a camel). This will launch a special-purpose text editor from which the user can create, run, and debug Perl scripts. Select New from the File menu to bring up a text-editing window. Type the script shown above, beginning with the #! /usr/bin/perl line and choose Save As from the File menu. Give the script a name, such as first .pl, and from the list of Type options in the Save dialog, choose Droplet. This will save the script as a miniature application called a *droplet*, which has a distinctive pyramid-shaped icon (containing another camel). Double clicking on this new icon will produce a text window containing the output of the script, which simply reads My first Perl script. The script can also be run directly through the MacPerl application by choosing Run Script from the Script pull-down menu.

How Script Work

A script consists of a series of commands, more formally called *statements*, that are meaningful to the Perl interpreter. Unless told otherwise, the interpreter starts at the top of the script file and works its way down to the bottom, executing each statement in turn.

Consider this new script:

```
#! /usr/bin/perl
# preamble...
print "I can do math! \n";
# do some calculations
$sum = 3 + 4;
# print the result
print "The sum of 3 + 4 is", $sum, ".\n";
```

This script consists of three statements; the first one (print "I can do math! \n") tells the interpreter to print out the indicated text. As we shall discuss in more detail later, the special character sequence \n is not interpreted literally but instead prints out as a *newline* character. The second statement, $sum = 3 + 4, adds the numbers three and four together and stores the result in a variable named $sum. The last statement prints the text The sum of 3+4 is, followed by the contents of the variable $sum, followed by a period and a newline.

Notice that each statement ends with a semicolon. The semicolon tells Perl where one statement ends and another begins. Blank lines and other white space can help make the script more readable but are ignored by the interpreter. Any line that begins with a pound sign (#) is a comment. When Perl sees a pound sign it simply ignores everything between it and the end of the line. The use of comments is strongly encouraged, since it allows other users to better understand what the programmer was trying to accomplish in a particular block of code, as well as reminding the programmer themselves of the same when reexamining code written long before.

The topmost line is also a Perl comment, but on UNIX systems it serves double-duty as a directive to the UNIX shell to tell it to execute the command /usr/bin/perl when the script file is executed. On Macintosh and Windows systems, this line is extraneous, but it is better to include it to maintain portability with UNIX machines and as a matter of good form.

Unless otherwise instructed, the Perl interpreter starts at the first statement and works its way to the last. When this script is run, the following output is produced:

```
I can do math!
The sum of 3+4 is 7.
```

Notice that the user only sees the output from the two print statements. The statement that performs the addition acts silently, behind the scenes.

Strings, Numbers, and Variables

Perl can deal with an astonishing number of data types, including but not limited to text, integers, floating point numbers, complex numbers, and binary numbers.

Following a long computer science tradition of using obscure terms for simple concepts, the Perl term for text is *string*. Strings are surrounded by single or double quotation marks:

```
'I am a string.'
"I am another string."
```

Having two types of quotation marks available makes it easier to create strings that contain embedded quotation marks:

```
'"Anna," she wailed "come quickly! The tiara is gone!"'
```

There are also some more substantial differences between the two types of quotes that will be discussed later in this chapter under *Variable Interpolation*.

Numbers are written just as one would expect:

```
1
49
28.2
-109
6.04E23
```

The last example shows how one represents scientific notation. The E means exponent, and the number should be interpreted as 6.02 10^{23}.

If you want your strings to contain special characters, such as tabs or new lines Perl provides special *escape sequences* to represent them. These escape sequences consist of a backslash followed by a single character. The two most commonly used are \n, which begins a new line, and \t, which inserts a tab. For example,

```
print "There is a newline\nhere and a tab\tthere. \n";
```

produces the following output:

```
There is a newline
here and a tab          there.
```

Perl only interprets escape sequences when they occur in double-quoted strings In single-quoted strings, the backslash and the character that follows it are interpreted literally.

Variables provide temporary storage for strings, numbers, and other values, li Perl, variables are arbitrary names preceded by a dollar sign. Examples of valid variable names are shown below.

```
$x
$X
$i_am_a_variable
$ LongVariableName
```

Perl variables are case sensitive. In the list above, $x is one variable, and $x is a different one entirely.

When first created, variables are empty or *undefined*. Values are assigned to variables using the = sign, also known as the assignment operator:

```
$x = 42;
print 'The value of $x is ',$x, "\n";
```

Assignment works from right to left. In the example above, the number 42 is assigned to the variable $x. Once assigned, the variable can be used in the place of a value, as the print statement above shows. The same variable can be used multiple times, using the assignment operator to change its contents.

```
print 'The value of $x is still ',$x, "\n",
$x = 'Mary had a little lamb';
print 'But now the value of $x is ',$x, "\n";
```

This code fragment will now print out the following:

```
The value of $x is still 42
But now the value of $x is Mary had a little lamb
```

Note that there is no restriction on the type of data a variable can hold. In this example, $x initially contained an integer and then held a string. Unlike some programming languages. Perl does not require the user to *declare* (formally describe) variables before using them, although this type of checking can be activated if desired. Perl actually has several types of variables. In addition to variables that hold a single value, which are technically called *scalar* variables, there are *arrays* and *hashes*, two types of variables that are capable of holding multiple values. These will be discussed in an upcoming example.

Arithmetic

Perl knows basic arithmetic. Symbols known as operators are responsible for the various arithmetic operations:

```
+  addition
—  subtraction
*  multiplication
/  division
** exponentiation
() grouping
```

The following example does a little math and prints out the result:

```
$x = 4;
$y=2;
$z = 3 + $x * $y?
print $z, "\n";
```

The result that gets printed out is 11. However, a more succinct way to express this would be to combine the first three lines into a single expression passed to print:

```
print 3+4*2, "\n";
```

Note that the arithmetic expression is processed as 3+(4*2) rather than (3+4)*2. When evaluating numeric expressions, Perl uses the standard rules of precedence. The precedence can be changed by explicitly using parentheses:

```
print (3+4)*2, "\n";
```

Variable Interpolation

Another interesting difference between double- and single-quoted strings is what happens when a variable is embedded inside a string. In double-quoted strings, the variable is expanded to its contents, a process known as *string interpolation*. This can aid readability considerably:

```
print "The value of $x is $x\n";
```

Assuming that $x again contains "Mary had a little lamb," the above statement outputs

```
The value of Mary had a little lamb is Mary had a little lamb
```

Single-quoted strings do not work in this fashion. If the print statement used a single-quoted string instead, it would print

```
The value of $x is $x\n
```

The user can precisely control whether variable interpolation occurs in double-quoted strings by placing a backslash in front of variables that should not be interpolated.

```
print "The value of \$x is $x\n";
```

In this statement, the first occurrence of the $x variable is protected against interpolation because of the backslash, but the second is not.

```
The value of $x is Mary had a little lamb
```

The backslash character can also be used to embed a double-quote character inside a double-quoted string. How is the following different from the outputs shown above? Run it through the Perl interpreter to check your conclusion:

```
print "The value of \$x is \"$x\"\n";
```

Variable interpolation only extends to the contents of the variable itself. Perl will not try to evaluate arithmetic expressions or other programming statements that are embedded in double-quoted strings. For example, the statements

```
$y= 19;
print "The result is $y+3\n";
```

produces the output

```
The result is 19+3
```

To evaluate the expression, put it outside the double quotes. Perl will do the arithmetic, and the print statement will output the result.

```
print "The result is ",$y+3,"\n";
```

BASIC INPUT AND OUTPUT

Input, in programming parlance, is how data "get into" a script. Output is, of course, what comes out of the script. Most scripts will do both, inputting data from one source and outputting it to another.

The main way of producing output is lo use the *print* function. Print takes a list of one or more arguments separated by commas and sends them to the current output device, which by default is the computer's screen. As already shown, print can deal equally well with text, numbers, and variables:

```
$sidekick = 100;
print "Maxwell Smart's sidekick is ", $sidekick-1, ".\n";
print "If she had a twin, her twin might be called"
2* ($sidekick-1), ".\n";
```

The result of this script is

```
Maxwell Smart's sidekick is 99.
If she had a twin, her twin might be called 198.
```

The main way to read input data is to use the angle bracket operator (< >), which reads a line of input from the current input device. You will usually call this operator in conjunction with the assignment operator to save the returned information into a variable:

```
$line = <>;
print "Got $line";
```

With these two operations, one can now write a fully interactive program named "dog years," which converts your age in human years to your age in dog years:

```
#! /usr/bin/perl
print "Enter your age: ";
$age = <>;
print "Your age in dog years is ", $age/7, "\n";
```

When this program is run, the result will look like this:

```
% dog—years .pi
Enter your age: 42
Your age in dog years is 6
```

where the 42 is typed in at the keyboard. Another tiny program illustrates one of the idiosyncrasies of the < > operator:

```
#! /usr/bin/perl
print "Enter your name: ";
$name = <>;
print "Hello $name, happy to meet you! \n";
```

Running this program produces output that might not be what is expected:

```
% hello.pl
Enter your name : Lincoln
Hello Lincoln
, happy to meet you!
```

What's going on? In fact, when the < > operator reads a line of input, the newline character at the end of the typed data is still there! More often than not, the newline character must be removed. Obligingly, Perl provides a function named *chomp* that will do exactly that, removing the terminal newline from a string. The rewritten program looks like this:

```
#!/usr/bin/perl
print "Enter your name: ";
$name = <>;
chomp $name;
print "Hello $name, happy to meet you! \n";
```

With the newline removed, the output looks the way it should:

```
% hello, pl
Enter your name: Lincoln
Hello Lincoln, happy to meet you!
```

The program can still be made a bit shorter by combining the input and chomp statements into a single statement, at the risk of making the program slightly harder to understand:

```
#! /usr/bin/perl
print "Enter your name: " ;
chomp ($ name = <>) ;
print "Hello $name, happy to meet you! \n" ;
```

The parentheses control the precedence of the operation so that Perl does the input first and then passes $name to the chomp function.

FILEHANDLES

When lines of input are read, the data come from the keyboard by default. When the script writes lines of output, the output goes to the screen by default. What if one wants to change these defaults so that input comes from a file or output goes into one?

There are several ways to do this, but the most straightforward is to use *filehandles*. A filehandle is the connection between a script and a file. Scripts can read from a filehandle to get the contents of the file a line at a time and print to a filehandle to add data to a file.

To open a file for reading, use the open function:

```
open MYFILE , 'data.txt';
```

The open function expects exactly two arguments. The first is a name for the filehandle (MYFILE). This is an arbitrary name that the user chooses, with the convention being to use all uppercase letters. The second argument is the name of the file to open. If just the file name is provided, as shown in the example, Perl will attempt to open a file by that name in the current directory (on Windows and UNIX systems, the directory in which the command to run the script was given: on Macintoshes, the folder that the script is located in). The full path, explicitly giving the location of the file, can also be given.

Unfortunately, the way one specifies a file path is different on the three different operating systems. On UNIX systems, a path begins with a forward slash and each directory is separated by additional slashes. On Windows systems, a path begins with the drive letter (e.g., C:) and uses backslashes to separate directories. On Macintoshes, the path begins with the name of the hard disk, and colons separate the name of each successive folder. Examples of fully qualified path names on UNIX. Windows, and Macintosh systems are shown below.

```
UNIX       /usr/local/blast/data/cosmids.txt
Windows    C:\Documents\Blast\Data\Cosmids.txt
Macintosh  HD:Sequence Data: Blast:cosmids
```

The open command may not be able to open a file for reading if, for instance, the file does not actually exist. Ways to detect and handle this kind of error will be discussed in the section *Filehandle Errors*.

Once a filehandle is open, it can be read from using the < > operator:

```
$line = <MYFILE>;
chomp $line;
$next_line = <MYFILE>;
chomp $next_line;
```

The only difference between reading from the keyboard and reading from a file is that, instead of using an empty pair of angle brackets (< >), the open filehandle is placed between them (<MYFILE>).

Each time you call <MYFILE>, a new line of data will be read from the file. When the last line has been read, the operation will return the undefined value; detecting this type of error will be discussed later. When a filehandle is no longer needed, it can be closed using the close function:

```
close MYFILE;
```

Writing to a file works in much the same way. The main difference is that, when the file is opened, Perl is instructed to write to the file by placing a > sign before the filename:

```
open MYFILE, '>data.txt';
```

If the file does not already exist, Perl will create it and open it for writing. If the file already exists, then Perl will empty out its existing contents before opening it. This ensures that the new data written to the file will replace anything that was already there. To add data to the end of the file without disrupting its current contents, the file can be opened for appending using the >> sign:

```
open MYFILE, '>>data.txt';
```

Data written to the file will now be appended to the end of its current contents rather than writing over the file. If the file does not already exist, then an empty one is automatically created.

Once a filehandle is opened for writing, data can be sent to it, using the filehandle as the print command's first argument:

```
print MYFILE "Your age in dog years is" ,$age/7, "\n";
```

This will write the indicated line of text to the file associated with MYFILE. Notice that there is no comma between the filehandle and the list of data arguments! The full, generalized syntax for print is

```
print [FILEHANDLE] $datal [ , $data2 [,$data3 ....]]
```

The square brackets mean an argument is optional. One or more spaces is used to separate the optional filehandle from the first data argument, and commas are used to separate the individual items to be written to the filehandle.

When finished writing to a filehandle, use the close function to close it as before. If the program ends without explicitly closing the filehandle, Perl will close the file automatically.

Nothing prevents a script from having multiple filehandles open at the same time. This odd little program will write the first two odd-numbered lines of input to the file odd.txt and even-numbered lines to the file `even.txt`:

```
#! /usr/bin/perl
open ODD, ">odd.txt";
open EVEN, ">even.txt";
$line = <>;
print ODD $line;
$line = <>;
print EVEM $line;
$line = <>
print ODD $line;
$line = <>
print EVEN $line;
close ODD;
close EVEN;
```

There is a certain amount of repetition in this script, for sake of clarity. This script will be revisited shortly, using a more elegant loop to perform the same function. Each time < > is called, a new line of input is obtained either from the keyboard or the file. What happens when the end of the file is reached? Because there aren't any more lines to read, < > simply returns the undefined value.

Making Decisions

So far, the programs that have been discussed have been very linear. The Perl interpreter starts at the first statement and works its way to the last, executing each of them along the way.

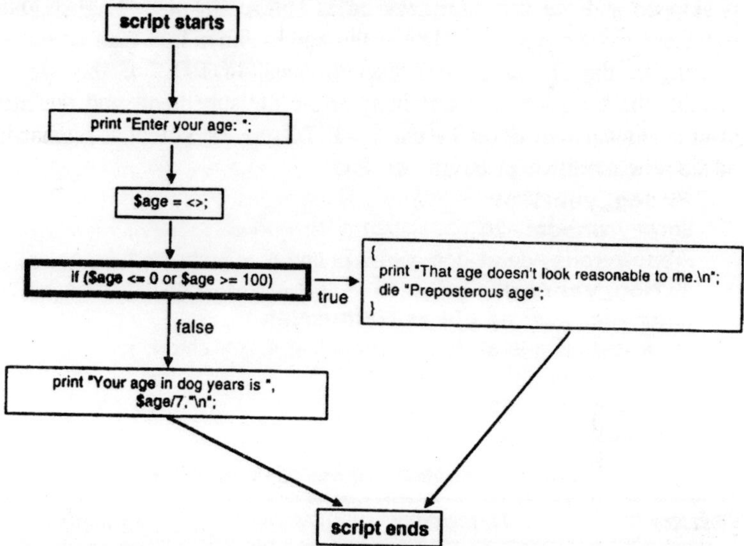

Fig. 5.1. The "dog years" calculator provides an example of entirely linear program flow.

Life is full of decisions, however, and so are Perl scripts. Often, there should be different paths followed if a particular condition is true, another if the condition is false. Considering the dog years calculator shown above, what would happen if the user had entered a negative number for the age or a number that's unreasonably large? It would be desirable to reject the input outright, rather than produce and output a preposterous answer. This simple modification to the original script achieves this goal.

```
#! /usr/bin/perl
print "Enter your age:";
$age = <>;
die "Preposterous age" if $age <= 0 or $age >=100;
print "Your age in dog years is", $age/7, "\n";
```

The fourth line is the new one, consisting of two parts. The first part (die "Preposterous age") introduces the *die* function, which is a lethal form of the print command. It prints out the error message given to it, along with information indicating the current line number. It then immediately terminates the program. It is great for handling those occasions when a fatal, unrecoverable error has occurred. The second part of the line (if $age <= 0 or $age >= 100) tests the $age variable's numeric value. If $age is either less than or equal to zero or greater than or equal to 100, then the test is true and the die function is executed. If $age satisfies neither test, then the die function is skipped and the program goes on as before. In plain English, the statement can be read as, "Die if the age is either less than or equal to zero or the age is greater than or equal to 100." If the age is invalid, the program will terminate at the die statement, and the last print statement will never be executed. Testing the potential scenarios of the now-modified program, we find

```
% dog_years.pl
Enter your age: -20
Preposterous age at dog_years.pl line 4, <> chunk 1.
% dog_years.pl
Enter your age: as old as Methuselah
Preposterous age at dog_years.pl line 4, <> chunk 1.
% dog_years.pl
Enter your age: 42
Your age in dog years is 6
```

Table 5.1. Numeric comparison operators

Operator	*Description*	*Example*
==	Equality	`$a == $b`
!=	Not equal	`$a != $b`
<	Less than	`$a < $b`
>	Greater than	`$a > $b`
<=	Less than or equal to	`$a <= $b`
>=	Greater than or equal to	`$a >= $b`
!	Logical not	`$ = ! $b`

Perl has a complete set of comparison operators that work on numbers and strings. To compare numbers, use any of the operators shown in Table 5.1. To compare strings, use any of the operators shown in Table 5.2. The numeric comparison operators are straightforward because they look, for the most part, like conventional expressions used in algebra. The big trap is the == operator that is used to test the equality of two numbers. Two equal signs are used instead of just one.

If you forget and accidentally use the assignment operator (=), then you will not get the result you expect:

```
$a == $b; # compare $a to $b, return true if equal
$a = $b; # assign contents of $b to $a
```

Read the ! operator as *not*. Unlike the other operators, it takes a single argument and reverses its truth. True expressions become false and vice versa. For example

```
print "the number is not greater than 0" if ! ($a > 0);
```

This statement first compares the current value of $a to zero and returns a true value if $a is greater than zero. The ! operator then reverses the test so that the expression as a whole is true only if $a is less than or equal to zero.

The string comparison operators are funny two-letter commands. The one used most frequently is eq, for testing whether two strings are the same. Consider this new version of the hello.pl script.

Table 5.2. String comparison operators

Operator	*Description*	*Example*
eq	Equality	$a eq $b
ne	Not equal	$a ne $b
lt	Less than	$a lt $b
gt	Greater than	$a gt $b
le	Less than or equal to	$a le $b
ge	Greater than or equal to	$a ge $b
=~	Pattern match	$a =~ /gattc/

```
# '/usr/bin/perl
print "Enter your name: ";
chomp ($name = <>);
print "Hello $name, happy to meet you! \n";
print "Hail great leader! \n" if $name eq 'Lincoln';
```

The output of the program uses eq to give Lincoln a special greeting:

```
% hello.pl
Enter your name: George
Hello George, happy to meet you!
% hello.pl
Enter your name: Lincoln
Hello Lincoln, happy to meet you!
Hail great leader!
```

Other handy string comparison operators are lt, which is true if the first string is less than the second string, and gt if the first is greater than the second. Perl compares strings alphabetically but uses criteria different from the telephone book. Among other subtle (and not-so-subtle) differences, the set of uppercase letters is less than the set of lowercase letters; in Perl, Z is less than a. Be careful not to use == to compare strings or eq to compare numbers. The handiest string comparison operator of them all is =~, the pattern matching operator. This is the most powerful of Perl operators, deserving its own discussion later in this chapter.

Two or more comparison operations can be combined using the operators and and or. An expression involving and is true only if *both* the right and left sides are true, whereas or expressions are true if *either* side is true. There is also a not operator, which reverses the sense of whatever comes to the right, making true expressions false and false ones true, just like Big Brother did in George Orwell's 1984.

The use of the if statement has already been demonstrated, executing a statement only when the condition that follows is true. As always with Perl, the opposite operator exists, called unless, executing a statement only when the condition following is false. Returning to the program testing for preposterous ages, the test could be rewritten this way:

```
die "Preposterous age" unless 0 < $age and $age < 100;
```

Read the statement this way: "Die unless the age is greater than zero and less than 100." The effect is the same. Sometimes it will seem more natural to write the conditional with an if sometimes with an unless.

Conditional Blocks

How would one approach executing several statements conditionally? In this case, the statements can be grouped into a block, using curly braces. The grouped statements can then be executed altogether, using the block form of if. Returning again to the dog-age calculator

```
#! /usr/bin/perl
print "Enter your age: ";
$age = <>;
if ($age <= 0 or $age >= 100) {
    print "That age doesn't look reasonable to me. \n";
    die "Preposterous age";
}
print "Your age in dog years is", $age/7, "\n"";
```

In this example, if controls a block of two statements surrounded by the curly braces. The first statement prints out a warning, and the second one terminates the program with die. If the age does not fall within the range of 0-100, then the statements within the block *will* execute and the program will end before reaching the last line of code. If the age does fall within the specified range, the two statements

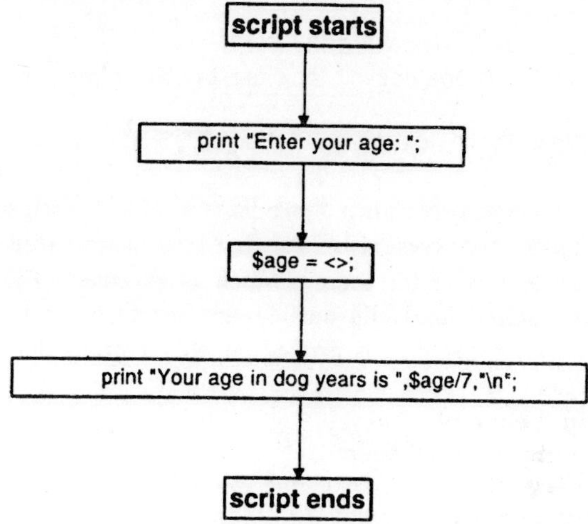

Fig. 5.2. If blocks can change the flow of execution.

in the if block are ignored and the program goes on to print out the calculated result, as before. The effect is to create two alternative paths in the program, one of which leads to termination with an error statement.

If blocks have this general form:

```
if (TEST) {
        STATEMENT 1;
        STATEMENT 2;
        STATEMENT 3;
             . . .
}
```

The test itself must be enclosed by parentheses in the manner shown, but any comparison operation involving numbers or strings is allowed. The indentation is a matter of style—although Perl does not depend on the indentation to interpret the code, indenting code in this fashion is invaluable when debugging a program, as it is easier to see where a given block of code begins and ends.

Perl can also handle situations in which one set of statements should be executed if the test condition is true, and another set of statements if the test condition is false. To do so, an else block is added to the if structure, changing the construct from an *if* statement to an *if—else* statement,

```
#! /usr/bin/perl
print "Enter your age:";
```

```
$age = <>;
if ($age <= 0 or $age >= 100) {
     print "That age doesn't look reasonable to me. \n";
} else {
print "Your age in dog years is" , $age/7, "\n" ;
}
```

The if-else statement shown above has two blocks, each surrounded by curly braces. The: contents of the first block is executed when the test is true. Otherwise the second Mock is executed. The result is that, if the entered age falls outside the acceptable range for the program, an error message is printed; if not, it prints the calculated results instead.

```
% dog_years.pl
Enter your age: eighteen
That age doesn't look reasonable to me.
% dog_years.pl
Enter your age: 21
Your age in dog years is 3
```

There happens to be a single statement in each block in this example, but there is no limit to the number of statements that can be enclosed within a block. Finally, for sake of completeness, code that specifies what is wrong with bad input data can be included in the program. A handy elsif block can test one string after another.

```
#! /usr/bin/perl
print "Enter your age:";
$age = <>;
if ($age <= 0) {
     print "You are way too young to be using a computer. \n" ;
} elsif ($age >= 100) {
     print "Not in a dog's life! \n";
} else {
print "Your age in dog years is", $age/7, "\n" ;
}
```

There are now two $age checks in this program. The first test compares the age with zero and prints out a warning message if it is less than or equal to zero. If the test is false, then the program proceeds to the elsif block and tries the second test, which compares $age with 100. If *this* test is true, then the second error message is printed out. Otherwise, if both tests are false, the program falls through to the else block. The possible outcomes are now as follows:

```
% dog_years.pl
Enter your age: -20
```

```
You are way too young to be using a computer.
% dog_years.pl
Enter your age: 999
Not in a dog's life!
% dog_years.pl
Enter your age: 28
Your age in dog years is 4
```

WHAT IS TRUTH?

The words *true* and *false* have been tossed around rather blithely in this chapter. It is now appropriate to define these terms more pricisely. Regardless of the meaning of truth in the broader, philosophical sense, truth in Perl boils down to four very simple rules:

Zero (0) is *false*.

"The empty string ("") is *false*.

The undefined value is *false*.

Everything else is *true*.

The various numeric and string comparison tests that were illustrated in the previous section evaluate to 1 when true and to the undefined value when false.

LOOPS

Conditional statements allow the flow of a program to be modified so that sections of code can be executed or skipped over as needed. They cannot, however, make the program execute a particular section of code more than once. For this, Perl (and most programming languages) utilizes what are known as *loops*. Perl has a lot of different types of loops available to the user, but the most useful one is the *while* loop. The while loop looks very much like an *if* block, but instead of executing the contents of the block once if the test is true, it executes the statements repeatedly so long as the test is true. An example of this type of loop is illustrated by this simple counting program:

```
#! /usr/bin/perl
$count = 1;
while ($count <= 5) {
     print "$count potato\n";
     $count = $count + 1;
}
```

Before the loop begins, the program creates a variable named $count and sets its value to 1. The while loop test checks whether $count is less than or equal to 10 and execute the two statements

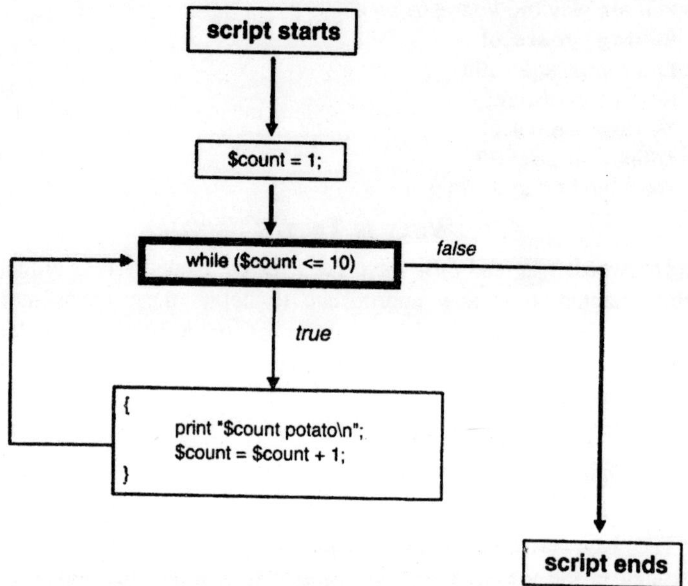

Fig. 5.3. While loops execute the same block repeatedly until a defined condition is satisfied.

contained in the curly braces as long as this condition holds true. The first statement prints out the current value of $count, and the second increments the variable by 1. The first time the while statement is encountered, $count was set to 1, so the block is executed. The second time, $count is 2, and the block executed again. This continues until $count is 6, at which point the test is no longer true (because $count is greater than 5) and the loop terminates. The output for the program is quite simply

```
% count.pl
1 potato
2 potato
3 potato
4 potato
5 potato
```

In the same way that unless reverses the sense of the if statement, until can be used as an alternative to while. Until loops will execute the contents of a block until a certain test becomes true. The counting program could then be rewritten as follows, to yield the same output:

```
# ' /usr/bin/perl
$count = 1;
until ($count > 5) {
```

```
        print "$count potato\n";
        $count = $count + 1;
    }
```

Combining Loops with Input

Loops become very powerful when combined with input statements. Consider this simple example:

```
#! /usr/bin/perl
print "Type something>";
while ( defined ($line = <> ) ) {
    chomp $line;
    print "You typed '$line' \n\n";
    print "Type something>";
}
```

This script prints out the prompt Type something> and immediately enters a while loop. The while loop here is a little different from the ones shown before. Instead of doing a comparison, the while loop's test contains the expression defined ($line = <>). This expression looks a bit bizarre but can be explained very simply. The first thing that happens is that a line of input is read and assigned to the variable $line. Then, $line is then passed to a new function named defined, which returns true if the contents of a variable are defined, false otherwise. Recall that < > reads lines from a file (or keyboard) until it reaches the end of the file, at which point it returns undefined. This test is telling the while loop to read input lines one at a time until the end of the file is reached. The parentheses ensure that defined is called after the line is read into $line, not before.

For each line read in this way, the while loop executes three statements. The first statement calls chomp to remove the newline character from the end of $line. The second statement uses variable interpolation to insert the input line into the string "You typed '$line' \n\n" and then prints the resulting string out. The last statement displays the prompt again. The result looks like this:

```
% echo.pl
Type something> hi there
You typed 'hi there'
Type something> this is something
You typed 'this is something'
Type something> ^D
%
```

As you can see, this program echoes back everything typed in, like a particularly annoying child. Even if nothing were typed in, the program would simply echo back an empty string; this is because the script would read a single newline character, which it would then remove with chomp, yielding an empty string. To stop the program, an *end-of-file* character has to be sent. On UNIX and Macintosh systems, this is done by typing control-D (written ^D for short). On Microsoft Windows systems, this is done by typing control-Z (^Z). An alternative is to kill the program by typing control-C (^C). This stops the program dead in its tracks.

Because it can be inconvenient to remember obscure control characters, the program can be made a bit friendlier by allowing the user to type quit to exit the loop:

```
# ! /usr/bin/perl
print "Typing something. 'quit' to finish>";
while (defined ($line=<>) ) {
chomp $line;
    last if $line eq 'quit';
    print "You typed '$line' \n\n";
    print "Type something>";
     }
print "goodbye! \n";
```

The main change is the line that immediately follows the chomp. This introduces a new function named last, which acts as a loop modifier. It is only allowed to occur within the body of a loop. When executed, last causes the script to exit the loop immediately, even though the loop test may still be true. This statement compares the contents of $line to the string quit. If they match, the last function is executed and the loop finishes. Running the program now, entering the word *quit* exits the loop so that the last line (which prints "goodbye!") can be executed.

Because the process of reading and processing incoming data a line at a time is so common, Perl provides a handy shortcut. If < > appears all alone in the test part of a while statement, Perl will read a line into an automatic variable with the odd-looking name $_ (which is read as "dollar sign underscore") and then call defined on your behalf. Furthermore, many text-processing functions operate on $_ by default, including chomp. Taking advantage of this shortcut, one can rewrite the previous example in this way:

```
#! /usr/bin/perl
print "Type something, 'quit' to finish>";
while (<>) {
    chomp;
```

```
    last if $_ eq 'quit';
    print "you typed '$_' \n\n";
    print "Type something>";
    }
print " goodbye!\n";
```

STANDARD INPUT AND OUTPUT

When a Perl script reads a line from <> and prints without using a specified filehandle, it instead uses two automatic filehandles called STDIN and STDOUT. The statements

```
print "The answer to life is...";
print STDOUT "The answer to life is. . .";
```

are exactly equivalent. In the same vein, the statements $line =<>; and $line = <STDIN>; are almost equivalent. The subtle differences between <> and <STDIN> will be discussed below.

The names STDIN and STDOUT are derived from *standard input* and *standard output*, an idea popularized by the UNIX operating system. Standard input and standard output are abstract files from which a script can accept input and send output, respectively. When a script is first launched, standard input corresponds to the keyboard and standard output corresponds to the computer screen. On Windows and UNIX systems, standard output appears in the command interpreter window. On Macs, the output appears in a small scrolling window that MacPerl creates specifically for this purpose. When a Perl script is launched, the user has the option of changing where standard input and output come from and go to. The user can also arrange for the standard output of one script to be sent to the standard input of another script in assembly-line fashion. This is actually a very powerful facility, but one that is beyond the scope of this chapter.

To redirect standard *input* from a file in either Microsoft Windows or UNIX systems, use the less-than (<) symbol to indicate the file:

```
% count_words.pl <C:\My Document\cosmids\11_22_00
cosmids.txt
```

To redirect standard *output* to a file on Windows or UNIX systems, use the greater-than (>) symbol:

```
% reverse_translate.pl > dna.txt
```

Standard input and output can be redirected simultaneously by using both symbols on the same command line:

```
% raverse_translate.pl <protein, txt> dna.txt
```

There is actually a third automatic filehandle called STDERR, for *standard error*. Perl sends its error messages and other diagnostics to standard error rather than standard output. When die is used to display error messages, the messages go to the STDERR filehandle automatically. STDERR is initially attached to the screen, just like STDOUT. On UNIX systems, STDERR can be redirected using >&. The following command will send the standard output of reverse_translate.pl to a file named dna.txt, and send any warnings or errors to a file named errors.out.

```
% reverse_translate.pl > dna.txt >& errors.out
```

Perl in Windows and Macintosh systems do not use the concept of a separate, standard error, so this type of redirection will not work on those systems.

Returning to the idea that <STDIN> and <> are almost—but not exactly—the same, the difference lies in the fact that <> contains some additional magic that makes it easy to process command line arguments. Given a file named unsorted.txt that is to be processed using a file named odd_even.pl, one way to process this file would be to redirect standard input, like so:

```
% odd_even.pl < unsorted.txt
```

However, because odd_even.pl contains a line using <> operator to read from standard input, the command can also be written in the following way:

```
% odd_even.pl unsorted.txt
```

The file name unsorted.txt is now being passed to the script as an *argument* and not as standard input. When <> is used, it looks for any filenames on the command line, opens them, and reads from them a line at a time. If no files are mentioned on the command line, <> reads from standard input. This feature can be used to process multiple files at once as well, reading from each of the files mentioned on the command line as is they were all a single, continuous file. Macintosh users, who do not have a command line available to them, can still take advantage of this <> feature. When a Perl script is saved as a droplet, text files can be dragged and dropped on top of its icon. Each of the dropped text files will be passed to <> for reading.

Finding the Length of a Sequence File

Moving to a biologically based example, consider a text file containing a large DNA sequence in single-letter format. The sequence is of unknown length, and it would be desirable to quickly determine the number of bases in the sequence. Using the file size alone would not be appropriate, since the presence of end-of-line characters at the end of each line would artificially inflate the number. The Perl script that will be developed in this section will answer this question as well as count up the number of lines in the original text file. This short script will read in the file, one line at a time, removing each line's terminal newline character and determining the length of what is left. The length for that line is added to a running total, with a counter tracking the number of lines in the file.

```
#! /usr/bin/perl
# file_size.pl
$length = 0;      # set length counter to zero
$lines = 0;       # set number of lines to zero
while (<>) {      # read file one line at a time
    chomp;  # remove terminal newline
     $length = $length + length $_;
    $lines = $lines + 1;
}
print "LENGTH = $length\n";
print "LINES = $lines\n";
```

A built-in function named length is invoked to determine the length of the line once the terminal newline is removed. When the script is done, it prints out the ultimate values of $length and $lines:

```
% file_size.pl dna.txt
LENGTH = 50649
LINES = 1387
```

Pattern Matching

The script from the previous section was useful only for calculating DNA lengths in the unusual case of having a file that contains nothing but raw DNA sequence. More frequently, however, sequence data come in FASTA format. To tally up the length of all the sequences in a FASTA file, the lines that begin with > must be ignored. Perl's pattern-matching operations make this easy to do.

A pattern match is a special type of text comparison. It is something like eq, but, instead of testing for an exact match between two strings, it tests a string against a pattern, using a pattern description language known as a *regular expression*. A simple example of a pattern match would be

```
print "EcoRI site found!" if $dna =~ /GAATTC/;
```

This if statement compares the contents of the variable $dna against the pattern GAATTC. The funny-looking =~ symbol is the pattern match comparison operator: think of it as an "approximately equal" comparison. If the string to the left of the pattern match operator contains the indicated pattern, it will return true. In the script fragment above, the program will print out EcoRI site found! if the string contains GAATTC anywhere along its length.

Regular expression patterns are delimited by forward slashes. The simplest ones contain a sequence of normal characters that must match somewhere within the body of a string. The *EcoR* I-site detector is one such example. Regular expressions are much more powerful than this, however. For example, square brackets can be used to specify a set of alternative characters in the manner shown here:

```
$dna =~ /GGG[GATC]CCC/
```

This pattern matches a sequence of characters beginning with GGG, followed by any of the characters G, A, T, or C, followed by the sequence CCC. In other words, this pattern searches for GGGNCCC.

To search for a series of alternative patterns, you can use the | symbol to separate the alternatives. For example, this will search for either *EcoR* I sites or *Hind* III sites:

```
$dna =~ /GAATTC|AAGCTT/;
```

This facility is greatly enhanced by *metacharacters* and *quantifiers*. A metacharacter represents a whole class of characters. For example, a single dot (.) will match any character except the end of a line, whereas \d signifies any digit. There are also metacharacters that will match the beginning and ending of lines and match the boundaries between one word and the next. Notice that there are many cases in which a metacharacter representing a character set is paired with its complement. For example, \s matches white space, and \s matches nonwhite space (in other words, printing characters). Another frequently used pair, ^ and $, match the beginning and end of lines, respectively.

Table 5.3. Regular expression metacharacters

Metacharacters	*Description*
.	Any character except newline
^	The beginning of a line
$	The end of a line
\w	Any word character (non-punctuation, non-white space)
\W	Any non-word character
\s	White space (spaces, tabs, carriage returns)
\S	Non-white space
\d	Any digit
\D	Any non-digit

For example, to match a "ZIP+4" format ZIP code, the required regular expression would be written as follows:

```
$address =~ /\d\d\d\d\d-\d\d\d\d/;
```

For a regular expression to match a metacharacter *literally*, it must be preceded by a backslash. For example, to match DNA sequence IDs of the form "M58200.2," where a dot is used literally, the regular expression should be written

```
$sequence_id=~ /\w+\.\d+/;
```

By default, any character or metacharacter in a regular expression matches exactly once. By placing a quantifier after the character, Perl can match a character a specific number of times or a range of times. The simplest type of quantifier is {M}, which tells Perl to match the pattern exactly M times. Using this notation, the ZIP+4 regular expression could be rewritten as

```
$address =~ /\d{5}-\d{4}/;
```

Similar quantifiers include the form {M,N}, which will match at least *M* times but no more than *N*, {M,}, which matches at least *M* times; and {,N}, which will match no more than *N* times. To bring the examples back into a biological context, the following expression will match the plant mcrBC methylation site Pu-C-X(40- 80)-Pu-C, in which the center of the recognition site can be anywhere from 40 to 80 nucleotides long.

```
$sequence =~ /[AG]C[GATC]{40,80}[AG]C/;
```

Although the curly braces can be used to describe any quantification, there are some short-cut metacharacters that are used for the frequent cases.

Parentheses can be used to group parts of a regular expression, and then a quantifier can be applied to the entire group. For example, this regular expression will match normal five-digit ZIP codes as well as the ZIP+4 form:

```
$address =~/\d{5}(-\d{4})?/;
```

The \d{5} part matches a digit repeated exactly five times. This is followed by an optional section containing -\d{4}, a hyphen followed by four digits. The optional section is completely surrounded by parentheses to group it, and the group is followed by a ? symbol, meaning that it can match at most once and possibly not at all. Parenthesized groups can also be used to extract portions of regular expressions.

With regular expressions, the DNA length calculator can be rewritten so that it correctly ignores lines that begin with the > sign. The modified program looks like this:

```
#! /usr/bin/perl
# file_size2 .pl
$length = 0;
$ lines = 0;
while (<>) {
chomp;
$length = $length + length $_ if $_ =~ /^[GATCNgatcn]+$/;
$lines = $lines + 1;
}
print "LENGTH = $length\n";
print "LINES = $lines\n";
```

The key modification to the original script is the insertion of a conditional test on the statement that tallies the length of each line of text. The script tests $_ for a pattern match with the regular expression /^[GATCNgatcn]+$/, which matches lines containing DNA sequence. The initial ^ character matches the beginning of the line, [GATCNgatcn]+ matches one or more of the characters GATCN or their lower-case equivalents, and the $ matches the end of the line. Other lines in the FASTA file, including blank lines and the description lines, are ignored. The file length that is printed out corresponds now to the sequence only.

Table 5.4. Regular expression quantifiers

Quantifier	*Description*
?	0 or 1 occurrence
+	1 or more occurrences
•	0 or more occurrences
{N,M}	Between N and M occurrences
{N,}	At least N occurrences
{,M}	No more than M occurrences

Before leaving this script, some syntactical tricks can be applied to make the script more concise. If a regular expression appears alone without a =~ operator, then Perl assumes that the variable to be tested for a pattern match is $_. The string length function also behaves this way, returning the length of $_ if no variable is explicitly specified. So, the length-tallying line can be rewritten as

```
$length = $length + length if /^[GATCNgatcn] + $/;
```

A second shortcut is to append an i flag at the end of the regular expression right after the second slash. This puts the regular expression into *case-insensitive* mode, and allows the statement to be written as

```
$length = $length + length if /^[GATCN] + $/i;
```

Finally, because adding a value to a variable and storing the sum back into the same variable is such a common operation. Perl provides the shortcut operator += (read as "plus equals"). This operator takes a numeric value on its right side, adds it to the contents of the variable on its left side, and stores the result back into the same variable, all in one graceful step. Taking advantage of this feature gives statements of the form

```
$length += length if /^[GATCN]] +$/i;
```

Putting all these shortcuts together gives the final version of the DNA length-tallying script:

```
#! /usr/bin/perl
# file_size3.pl
$length = 0;
$lines = 0;
while (<>) {
    chomp;
    $length += length if /^[GATCN]+$/i;
    $lines += 1;
}
print "LENGTH = $length\n";
print "LINES = $lines\n";
```

Table 5.5. Assignment shortcut operators

Operator	*Example*	*Description*
+=	`$a += 3`	Add a number to a variable
-=	`$a -= 3`	Subtract a number from a variable
*=	`$a *= 10`	Multiply variable by a number
/=	`$a /=2`	Divide a variable by a number
.=	`$txt .="abc"`	Append a string to a variable

EXTRACTING PATTERNS

Not only are regular expressions good for detecting patterns in text, but they can be used to extract matching portions of the text as well. To see how this might be useful, consider a FASTA description line like this one:

```
>M18580 Clone 305A4, complete sequence
```

In addition to the initial >, the description line contains two different fields that we might like to capture. The first is "M18580," the mandatory sequence ID. The second is "Clone 305A4, complete sequence," an optional human-readable comment. In regular expression terms, the description line looks like this:

```
/^>\S+\s*.*$/
```

Reading from left to right is the beginning of the line (^), a > sign, one or more non-white-space characters corresponding to the sequence ID, zero or more spaces or other white space, zero or more of any character corresponding to the optional description, and the end of the line ($). This regular expression match can be made to extract the ID and description lines simply by putting parentheses around the parts that should be captured:

```
/^> (\S+)\s* (.*)$/;
$id= $1;
$description = $2;
```

When a string successfully matches a regular expression, any portions of the expression that are contained within parentheses are extracted and placed into the automatic variables $1, $2, $3, and so forth. The extraction works from left to right and only happens if the entire regular expression matches.

This trick can be used to fix a deficiency in the DNA length calculator from the previous section. Previous versions of this script naively treated the entire FASTA file as a single DNA sequence. However, a FASTA file usually contains entries for multiple sequences. With pattern matching, the description lines can be identified and the sequence IDs extracted, allowing for the length of each sequence to be printed out separately.

```
#! /usr/bin/perl
$id = '';                # holds sequence ID of current sequence
$length=0;               # holds length of current sequence
$total_length=0;         # tallies aggregate length of all seqs
while (<>) {
    chomp ;
    if (/^>(\S+)$/) {# found a new description line
      print "$id: $length\n" if $length > 0;
      $id= $1;
      $length = 0;
    } else {
      $length += length,
      $total—length += length;
    }
}
print "$id: $length\n" if $length > 0; # last entry
print "TOTAL LENGTH = $total_length\n";
```

After initializing the three variables at the top of the program, the script enters a while loop. As before, it reads a line at a time into $_ and removes the terminating newline. The new feature is an if-else block. The block performs a pattern match on the line, looking for a FASTA description line. If one is found, it signals the beginning of a new sequence. The following then occurs:

- If `$length` is non-zero, the ID and length of the previous sequence are printed. The check on `$length` prevents the program from printing out an empty line when it hits the very first sequence in the file.
- The matched sequence ID is copied into $id.
- `$length` is set to zero.

Otherwise, the program is in the middle of a sequence, in which case the length of the current line is added to the appropriate counters. After the last line is read, the ID and length of the last sequence in the file are printed, as well as the total length of all the sequences:

```
% fasta_length.pl ests.fasta
D28205: 1105
BCD207F: 402
BCD207R: 332
BCD386F: 192
BCD386R: 362
CD098F: 374
TOTAL LENGTH = 2767
```

Arrays

Previous examples have worked with single-valued *scalar* variables only. However, Perl has the ability to work with multivalued variables as well. There are two basic multivalued variables, named *arrays* and *hashes*. An array is a list of data values indexed by number. A hash is a list of data values indexed by string. Both are very easy to use and incredibly handy. To understand arrays, consider how one might keep track of a large number of identifiers, such as clone names. With scalar variables, one approach could be to assign each clone name to a different variable:

```
$clone1 = '192a8' ;
$clone2 = '18c10' ;
$clone3 = '327h1';
. . .
```

The problem with this approach, besides being tedious, is that it does not offer any way to step through the entire list of clones one by one, performing the same operation on each one. Arrays circumvent this problem. An array can be defined as follows:

```
@clones = ('192a8' , '18c10' , '327h1' , '201e4');
```

This new array, named @clones, contains four strings. Array variables begin with an @ sign to distinguish them from scalar variables, which begin with a $. Scalars and array variables are completely separate. In fact, a scalar variable can be named @clones and an array variable can be named @clones within the same program. They will not interact in any way. As alluded to above, operations can be applied to arrays as a whole. For example, the = operator can be used to copy one array into another:

```
$old_clones = @clones;
```

Items can be added to the end of an array using the *push* function:

```
push @clones, '281e3 ';
```

After this statement executes, @clones will contain five items. The opposite of push is *pop*, which removes the last item from the array, reducing it in size by one, and returns the removed item as its result. This statement will reduce @clones back to a length of four , assigning 281e3 to the scalar variable $last_clone:

```
$last_clone = pop @clones;
```

Two array operations are particularly common: accessing an arbitrary array element by its positions in the array using *indexing* and looping over each element of the array in order using the *foreach* loop. Considering indexing first, to copy the third element of @clones into a variable named $third_clone, the statement would be written as

```
$third_clone = $clones [2];
```

This will—and should—look strange at first. The numeral in the square brackets, [2], is the *index*. Perl numbers its arrays starting with *zero*, so the first item is actually index 0, the second item is index 1, and the third item is index 2. Any expression can be placed within the square brackets, as long as that expression ultimately evaluates to an integer. As an example, if a scalar variable $i contains a number, to address the next element in the series, it could be referred to as $clones [$i+1]. More mystifying, however, is the $ at the beginning of the array name. What happened to @clones?

When one indexes into an array, the symbol at the front refers to the individual array element, *not* the array as a whole. Because the element itself is a scalar, the symbol at the front should be a $. To clarify, look at the following two examples.

```
@old_clones = @clones;
$first_clone = $clones[0];
```

The first line would copy the *entire* array, whereas the second line copies just a single element within the array.

Arrays can be extremely long; ones with thousands of elements are not unusual. A common operation is to loop through each member of an array and do something with it. The foreach loop makes this possible. For instance, say that the array @dna contains a list of DNA sequences and that a printout of the length of each element would be helpful. The following small loop would accomplish this.

```
foreach $dna (@dna) {
     print length $dna, "\n";
}
```

The foreach loop has three parts: the name of a scalar variable known as the *loop variable*, an array name enclosed in parentheses, and a block containing the statements to be executed. Foreach steps through the array one element at a time, placing each element in the loop variable and executing the statements within the block. The statements may examine the contents of the loop variable and act on it or even change the loop variable to change the corresponding array element. After the loop is finished, the loop variable will again contain whatever it had before the loop began or be undefined if this is the only time it was used.

To illustrate how assigning to the loop variable changes the contents of the array, below is a fragment of code that will treat every element of an array as a DNA sequence, replacing it with its reverse complement:

```
foreach $dna (@dna) {
     $dna = reverse $dna; # reverse it
     $dna =~ tr/gatcGATC/ctagCTAG/; # complement it
}
```

The two statements in the block show off a pair of Perl functions that have not yet been discussed. The first of these, *reverse*, returns the reverse of a scalar variable, turning GGGGTTTT into TTTTGGGG. The reversed sequence is assigned back into the loop variable. The next statement uses the *tr* function (for *translate*) to substitute one set of characters with another. tr has an unusual syntax that is rooted in Perl's historical origins. It uses the slash as a delimiter, replacing the list of characters between the first set of delimiters with the characters in the second set. The replacement occurs on whatever variable tr is bound to using the =~ operator (the syntax should be reminiscent of pattern matching). Characters not mentioned in the list are left unchanged. In the example above, the list gatcGATC is replaced with ctagCTAG. What happens is that "g" is replaced with "c," "a" with "t", and so forth. Thus TTTTGGGG becomes AAAACCCC, which is the reverse complement of the original element, GGGGTTTT.

Arrays and Lists

Perl lists are closely related to arrays. Lists are a set of constants or variables enclosed in parentheses. An example of a list of strings would be ("one", "two", "buckle my shoe"), whereas an example of a list of variables, would be ($a, $b, $c). A list that combines variables, constant strings, and constant numbers might be something like ($a, "the Roman empire", 3.1415926, $ipath). Lists can be thought of as being related to array variables in the same way that the constant 123.4 might be related to the scalar variable $total.

Lists are useful for performing operations in parallel. For example, lists can be assigned to array variables to make the array identical to the list. In fact, one example of this was shown earlier:

```
@clones = ('192a8', '18c10', '327h1', '201e4');
```

Arrays can be assigned to lists, provided that each element of the list is a variable and not a constant. For example, to extract the first three elements of an array, one could write

```
($first, $second, $third) = @clones;
```

After this operation, $first will contain 18c10, and so on. Naturally enough, lists can be assigned to lists as well, again provided that the list on the left contains variables only:

```
($one, $two, $three) = (1,2,3);
```

Split and Join

It is often very useful to transform strings into arrays and to join the elements of arrays together into strings. The *split* and *join* functions allow for these operations. Split takes two arguments: a delimiter and a string. It splits the string at delimiter boundaries, returning an array consisting of the split elements. The delimiters themselves are discarded. To illustrate this, consider a case requiring the manipulation of a long file containing comma-delimited files, such as the following:

```
192a8, The Sanger Centre, GGGTTCCGATTTCCAA,
CCTTAGGCCAA ATTAAGGCC
```

Split makes it easy to convert the long string into a more manageable array. To split on the comma, the comma is used as the delimiter:

```
chomp($line = <>);          # read the line into $line
@fields = split ',', $line;
```

@fields will now contain the five individual elements, which can now be indexed or looped over. Split is often used with a list on the left side instead of an array, allowing one to go directly to assigning to the list. For example, rather than creating an array named @fields, the result of a split command can assign values to a list of named, scalar variables:

```
($clone, $laboratory, $left—oligo, $right_oligo) = split ',', $line;
```

The join function has exactly the opposite effect of split, taking a delimiter and an array (or list) and returning a scalar containing each of the elements joined together by the delimiter. Thus, continuing the earlier example, the @fields array can be turned into a tab-delimited string by joining on the tab character (whose escape symbol is \t):

```
$tab_line = join "\t", @fields;
```

After this operation, $tab_line will look like this:

```
192a8 The Sanger Centre GGGTTCCGATTTCCAACCTTAGGCC
AAATTAAGGCC
```

Hases

The last Perl data type that will be considered is the *hash*. Hashes are similar to arrays ill many respects. They hold multiple values, they can be indexed, and they can be looped over, one element at a time. What distinguishes hashes from arrays is that the elements of a hash are unordered, and the indexes are not numbers but strings. A few examples will clarify this:

```
%oligos = ( ) ;
$oligos{'192a8') = 'GGGTTCCGATTTCCAA';
$oligos{'18c10'} = 'CTCTCTCTAGAGAGAGCCCC';
$oligos{'327h1'} = 'GGACCTAACCTATTGGC';
```

In this example, an empty hash named %oligos is created, and three elements are' then added to it. Each element has an index named after the clone from which it was derived, and a value containing the sequence of the oligo itself. After the assignments, the values can be accessed by indexing into the hash with curly braces.

```
$s = $oligos{'192a8'} ;
print "oligo 192a8 is $\n";
print "oligo 192a8 is", length $oligos{'192a8'}, " base pairs long\n"
;
print "oligo 18c10 is $oligos{'18c10'}\n";
```

This will print out

```
oligo 192a8 is GGGTTCCGATTTCCAA
oligo 192a8 is 16 base pairs long
oligo 18c10 is CTCTCTCTAGAGAGAGCCCC
```

Just as a variable containing an integer can be used as an index into an array, a variable containing a string can be used as the index into a hash. The following example uses a loop to print out the sequence of each of the three oligos previously defined by %oligos:

```
foreach $clone ('327h1', '192a8', '18c10') {
      print "$clone: $oligos{$clone}\n";
}
```

As with arrays, there is a distinction between the hash as a whole and individual elements of a hash. When referring to an element of a hash using its index surrounded by curly braces, one is referring to the *scalar* value contained within the hash, so the $ symbol must be used as a prefix. To refer to the hash variable as a whole, use the % symbol as-the prefix. This allows for one hash to be assigned to another, as well as the ability to perform other whole-hash operations.

For historical reasons, the indexes of a hash are called its *keys*. Calling the *keys* function produces a list of all the keys in the hash. Using a command of the form @clones = keys %clones; will assign to the array three string elements ('327h1','192a8','18c10'), *but in no predictable order*. The elements of a hash are unordered, and the order in which they are put into a hash has no effect on the order in which they are returned.

To get all the values of a hash, use the *values* function:

```
@oligos = values %clones;
```

The @oligos array will now contain a three-element list consisting of each of the oligo sequences that were placed into the hash. As with keys, the values are returned in an unpredictable order. However, the order of elements retrieved by keys will match the order retrieved by values. Hence, the position of clone 192a8 in the @clones array will match the position of its corresponding oligo in @oligos.

A Real-World Example

At this point in the discussion, all the tools needed to solve the problem posed at the very beginning of this chapter are now at hand. The problem will be approached in steps. The first task is to scan through a file of RNAi results, collecting the genes that have anything to do with locomotion. Assume that the input file is named rnai.txt and contains lines in the following format; each field is separated by a tab.

The main challenge here is to search the Phenotype Summary field for results having to do with locomotion. This is a fuzzy sort of problem, best solved using Perl's pattern-matching facility. After scanning through the results for a while, we decide to search for any of the following keywords: uncoordinated, paralysis, paralyzed, coils, coiler, movement, kinky, and jerky.

With this decision made, the first part of the script can be written. The script attempts to open the file rnai.txt. If unsuccessful, it dies with an error message. If successful, it initializes the hash %interesting_genes to empty. This hash will be used to hold the list of locomotion-related genes and will be set up so that the keys are gene names and the values are the corresponding phenotypes. The program then steps through the input file, one line at a time. It chomps off the newline from the end of each line and then uses the split function to split each line into fields, using the tab character as the delimiter. This results in a four-element list, which is assigned to an array containing the variables $gene, $date, $status, and $phenotype.

Once this is done, the contents of $phenotype are compared to a regular expression. The regular expression is a series of alternatives, separated by the | character. Any phenotype that contains any of the strings listed will produce a match. Notice that some of the keywords have been shortened to reduce the length of the expression, as well as pull in some terms that may not have been anticipated. For example, paraly will match both paralysis and paralyzed, without much risk of matching something unintended. If the pattern matches, then the corresponding gene is recorded. By the end of the loop, %interesting_genes will be populated with all of the genes whose phenotypes matched the target list. Because rnai.txt will no longer be used, its filehandle is closed.

In Step 2, the WormPep set of predicted *C. elegans* proteins will be stepped through, pulling out all the ones matching the collection of genes identified in Step 1. WormPep's format is similar to the standard FASTA format:

```
>2L52.1 CE20433 Zinc finger, C2H2 type (CAMBRIDGE) protein
id:CAA21776.1
MSMVRNVSNQSEKLEILSCKWVGCLKSTEVFKTVEKLLDHVTADHIPEVIV
NDDGSEEVVCQWDCCEMGASRGNLQKKKEWMENHFKTRHVRKAKIFKCL
IEDCPVVKSSSQEIETH...
```

The portion of the definition line that immediately follows the > is the name of the gene. The program needs to step through all of these entries, extracting the gene names and saving the sequences of those contained in the collection of interesting genes. The script begins Step 2 by opening the WormPep file using a filehandle named WP. If successful, it initializes a hash named %sequences. This hash will have keys corresponding to the names of the interesting genes, with values consisting of their peptide sequences. The script then enters a loop in which it retrieves each line of the WormPep file, pattern-matching it against a regular expression that examines the def lines. If the program detects a definition line, meaning the beginning of a new sequence, it puts the gene name into a scalar variable named $id, taking advantage of Perl's ability to extract parenthesized portions of regular expressions.

If the current line does *not* match the regular expression, then the program has hit a sequence line, with the name of the current gene being held in $id. The if statement tests whether the current gene is an element of %interesting_genes, and, if so, the sequence will be read in, growing one line at a time, until the next definition line is reached. At the end of the loop, %interesting_genes will be fully populated with the sequences of interest. Note that the newline has not been removed from the end of each line of input sequence data; in this case, it is desirable to keep the newlines, to facilitate later parts of the program.

In Step 3, the gene sequences are formatted into an E-mail message for submission to the Signal? server. The server expects E-mail submissions in the following format:

```
euk
graphics
>ID1 Comments (ignored)
MLETLCYNYLPLCEQLEPVLNVRDKEDLATSLVRVMYKHNLAKEFLCDLIMKEVEKL...
>ID2 More comments (ignored)
MPARRHLSQPAREGSLRACRSHESLLSSAHSTHMIELNEDNRLHPVHPSIFEVPNCF...
```

The first two lines contain information required by SignalP to properly process the sequences; here, the server is being instructed that the sequences are from a eukaryote and that graphics of the predictions should be returned. Following these flags are the sequences, in FASTA format. The sequence ID is required, and anything following it is ignored by the server. A dot follows the last sequence in the file.

The script begins Step 3 by attempting to open the file signalP.txt for writing. If successful, it writes the top two lines of the outgoing E-mail message to the file. The program then enters a foreach loop, calling the keys function to recover all keys from the %sequences hash. This retrieves all of the names of all of the genes for which sequence information has been assembled. For each gene, a description line containing the gene name and its phenotype is printed to the filehandle. Although the phenotype will be ignored by SignalP, the information is retained for future reference. After this, the sequence of the gene is printed, newlines and all. At the end of the loop, the signalP.txt filehandle is closed. The final step in the analysis is for the researcher to take the newly created signalP. txt file and to e-mail it to the SignalP server. The most useful part of this script is that it can be automatically rerun each time WormPep is updated to repeat the analysis.

WHERE TO GO FROM HERE

Perl has many features that cannot be covered in a single chapter. For example, subroutines allow one to define customized functions that take arguments and return a result. *References* allow sophisticated data structures such as lists of lists to be created. *Objects* allow large, complex programs to be written so that code can be reused in different contexts. *Pipes and processes* allow for the control of external programs, perhaps to create an automated pipeline invoking commonly used programs.

Last, but not nearly least, there are *modules*, which are libraries of useful code routines put together by Perl programmers around the world and made available for public use. For example, the Mail::Sendmail module would enable the SignalP-processing script to E-mail its formatted message directly to the SignalP server without creating an intermediate file. Other modules allow for the creation of interactive, graphical front ends for programs or the creation of dynamic Web pages. Most relevant for biologists are the modules that form BioPerl, an extremely powerful collection of tools for searching biological databases, manipulating and processing sequences, and analyzing nucleotides and proteins.

6

Prediction of RNA Secondary Structure

The methods used either align entire sequences or search for common patterns in the sequences. In either case, the objective is to locate a set of sequence characters in the same order in the sequences. Nucleic acid sequences that specify RNA molecules have to be compared differently. Sequence variations in RNA sequences maintain base-pairing patterns that give rise to double-stranded regions (secondary structure) in the molecule. Thus, alignments of two sequences that specify the same RNA molecules will show covariation at interacting base-pair positions. In addition to these covariable positions, sequences of RNA-specifying genes may also have rows of similar sequence characters that reflect the common ancestry of the genes.

Introduction

As genomic sequences of organisms become available, it is important to be able to identify the various classes of genes, including the major class of genes that encodes RNA molecules. There are a large number of Web sites that provide programs and guest sites for RNA analysis or for access to databases of RNA molecules and sequences. These molecules perform a variety of important biochemical functions, including translation; RNA splicing, processing, and editing; and cellular localization. As with proteins, RNA-specifying genes may be identified by using the unknown gene as a query sequence for DNA sequence similarity searches. If a significant match to the sequence of an RNA molecule of known structure and function is found, then the query molecule should have a similar role. For some small molecules,

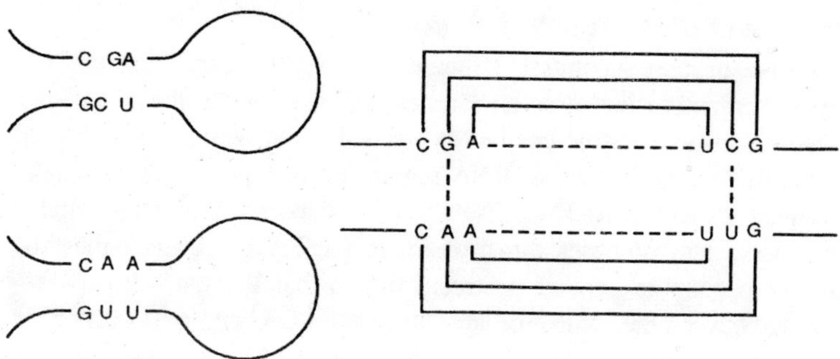

Fig. 6.1. Complementary sequences in RNA molecules maintain RNA secondary structure.

the amount of sequence variation necessitates the use of more complex search methods, described later in this chapter.

RNA Structure Prediction Basics

A computational method for predicting the most likely regions of base-pairing in an RNA molecule has been designed, just given the sequence, thus providing an ab initio prediction of secondary structure. From the many possible choices of complementary sequences that can potentially base-pair, the compatible sets that provide the most energetically stable molecules are chosen. Structures with energies almost as stable as the most stable one may also be produced, and regions whose predictions are the most reliable can be identified from such an analysis. Sequence variations found in related sequences may also be used to predict which base pairs are likely to be found in each of the molecules. One variation of RNA structure prediction methods will predict a set of sequences that are able to form a particular structure. Methods for predicting three-dimensional structures from sequence are also being developed.

Another type of RNA secondary structure prediction method takes into account conserved patterns of base-pairing that are conserved during evolution of a given class of RNA molecules. Sequence positions that base-pair are found to vary at the same time during evolution of RNA molecules so that structural integrity is maintained. For example, if two positions G and C form a base pair in a given type of molecule, then sequences that have C and G reversed, or A and U or U and A at the corresponding positions, would be considered reasonable matches. These patterns of covariation in RNA molecules are a manifestation of secondary structure that lead to a structural prediction. The computational challenge is to discover these covariable positions against the background of other sequence changes.

Features of RNA Secondary Structure

Like protein secondary structure, RNA secondary structure can be conveniently viewed as an intermediate step in the formation of a three-dimensional structure. RNA secondary structure is composed primarily of double-stranded RNA regions formed by folding the single-stranded molecule back on itself. To produce such double-stranded regions, a run of bases downstream in the RNA sequence must be complementary to another upstream run so that Watson—Crick base-pairing between the complementary nucleotides G/C and A/U (analogous to the G/C and A/T base pairs in DNA) can occur. In addition, however, G/U wobble pairs may be produced in these double-stranded regions. As in DNA, the G/C base pairs contribute the greatest energetic stability to the molecule, with A/U base pairs contributing less stability than G/C, and G/U wobble base pairs contributing the least. From the RNA structures that have been solved, these base

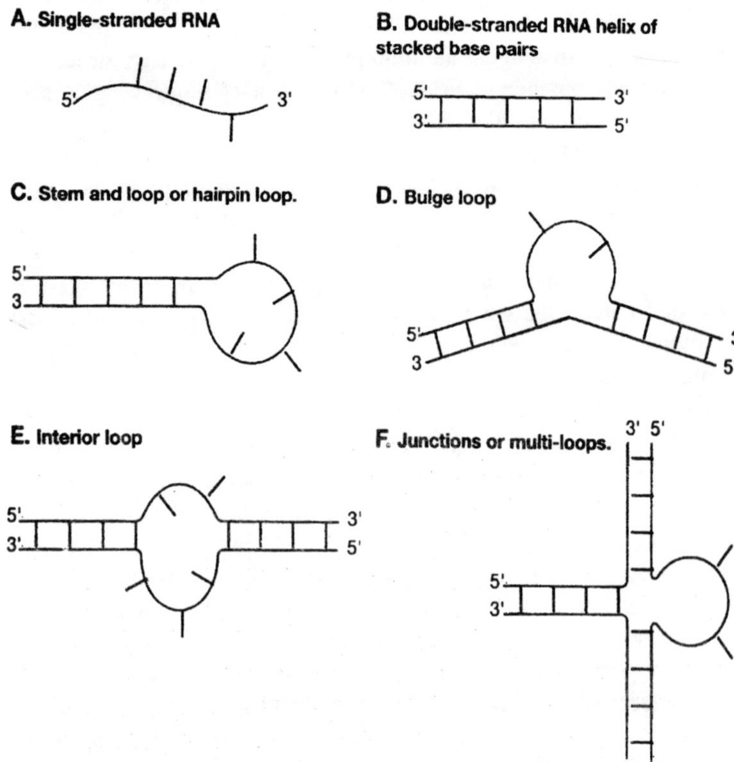

Fig. 6.2. Types of single- and double-stranded regions in RNA secondary structures.

pairs and a number of additional ones have been identified. RNA structure predictions comprise base-paired and non-base-paired regions in various types of loop and junction arrangements.

In addition to secondary structural interactions in RNA, there are also tertiary interactions. These kinds of structures are not predictable by secondary structure prediction programs. They can be found by careful covariance analysis.

Limitations of Prediction

In predicting RNA secondary structure, some simplifying assumptions are usually made. First, the most likely structure is similar to the energetically most stable structure. Second, the energy associated with any position in the structure is only influenced by local sequence and structure. Thus, the energy associated with a particular base pair in a double-stranded region is assumed to be influenced only by the previous base pair and not by the base pairs farther down the double-stranded region or anywhere else in the structure. These energies can be reliably estimated by experimentation with small, synthetic RNA oligonucleotides recently improved to include sequence dependence. They are most reliable when used for standard Watson-Crick base pairs and single G-U pairs surrounded by Watson-Crick pairs. Finally, the structure is assumed to be formed by folding of the chain back on itself in a manner that does not produce any knots. The best way of

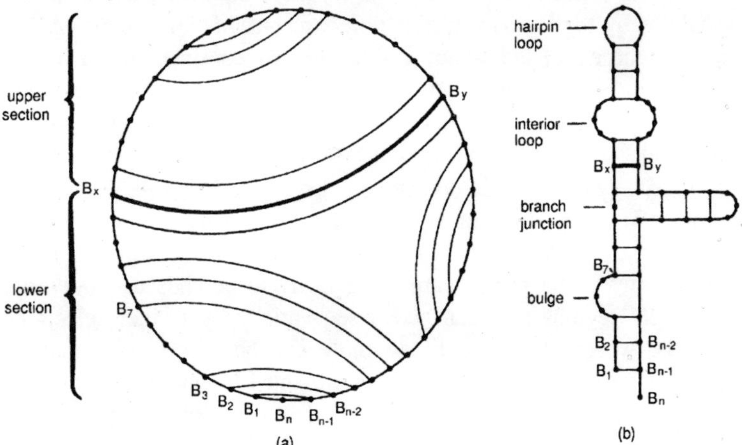

Fig. 6.3. Display of base pairs in an RNA secondary structure by a circle plot. The predicted minimum free-energy structure shown in (b) is represented by a plot of the predicted base pairs as arcs connecting the bass in the sequence, which is drawn around the circumference of a circle, as shown in A.

representing this requirement is to draw the sequence in a circular form. The paired bases are then joined by arcs. If the total structure with all predicted base pairs is to be free of knots, none of the arcs must cross. Note, however, that if a pseudoknot is represented on such a diagram, the lines will cross.

Development of RNA Prediction Methods

The development of methods for predicting RNA secondary structure has been reviewed by von Heijne (1987). Tinoco et al. (1971) first estimated the energy associated with regions of secondary structure by extrapolation from studies with small molecules and then attempted to predict which configurations of larger molecules were the most energetically stable. Energy estimates included the stabilizing energy associated with stacking base pairs in a double-stranded region and the destabilizing influence of regions that were not paired. Pipas and McMahon (1975) developed computer programs that listed all possible helical regions in tRNA sequences; using modified Watson-Crick base-pairing rules, they created all possible secondary structures by forming permutations of compatible helical regions, and evaluated each possible structure for total free energy. Studnicka et al. (1978) designed a method for adding compatible double-stranded regions together to produce the energetically most favourable structure. Martinez (1984) made a list of possible double-stranded regions, and these regions were then given weights in proportion to their equilibrium constants, calculated by the Boltzmann function [exp $(-\Delta G/RT)$], where $-\Delta G$ is the free energy of the regions, R is the gas constant, and T is the temperature. The RNA molecule is folded by a Monte Carlo method in which one initial region is chosen at random from a weighted pool, similar to the method used in Gibbs sampling.

Imagine each possible double-stranded region being represented by a marble in a bag. The number of each type of marble is weighted by the Boltzmann probability so that marbles corresponding to more energetically stable regions are more likely to be chosen. Additional compatible regions are then added sequentially by further selections from the weighted pool until no more can be added. This method generates a set of possible structures weighted by energy, but it does not take into account the destabilizing effect of unpaired regions. The Boltzmann probability function is used in more recent applications to find the most probable secondary structures.

Nussinov and Jacobson (1980) were the first to design a precise and efficient algorithm for predicting secondary structure. The algorithm generates two scoring matrices—one $M(i,j)$ to keep track of the maximum number of base pairs that can be formed in any interval i to j in the sequence and a second $K(i,j)$ to keep track of the base position k that is paired with j. From these matrices, a structure with the maximum possible number of base pairs could be deduced by a trace-back procedure similar to that used in performing sequence alignments by dynamic programming. Zuker and Stiegler (1981) used the dynamic programming algorithm and energy rules for producing the most energetically favourable structure. Their method assumes that the most energetic, and usually longest, predicted dsRNA regions are present in the molecule. Because many double-stranded regions are predictable for most RNA sequences, the number of predictions is reduced by including known biochemical or structural information to indicate which bases should be paired or not paired, by enforcing topological restraints and by requiring that the structure be in an energetically stable configuration.

MFOLD, written by Dr. Michael Zuker and colleagues, is commonly used to predict the energetically most stable structures of an RNA molecule. MFOLD provides a set of possible structures within a given energy range and provides an indication of their reliability. The program also uses covariance information from phylogenetically related sequences. MFOLD includes methods for graphic display of the predicted molecules. This program is one of the most demanding on computer resources that is currently used because the algorithm is of N^3 complexity, where N is the sequence length. For each doubling of sequence length, the time taken to compute a structure increases eightfold. The program also requires a large amount of memory for storing intermediate calculations of structure energies in multiple scoring matrices. As a result, MFOLD is most often used to predict the structure of sequences less than 1000 nucleotides in length. This method is most reliable for small molecules and becomes less reliable as the length of the molecule increases.

MFOLD and many other types of useful information on RNA are found at the Web site of Dr. Michael Zuker, at http://bioinfo.math.rpi.edu/~zuker/rna/. Details of running MFOLD are not given here because the user manual for MFOLD is widely available. Recently, a new method called the partition function method for finding the most probable secondary structural configuration of an RNA molecule and the most probable base pairs has been reported by the Vienna RNA group.

One advance in the prediction of RNA structure has come from the recognition that certain RNA sequences form specific structures and that the presence of these sequences is strongly predictive of such a structure. For example, the hairpin CUUCGG occurs in different genetic contexts and forms a very stable structure. Databases of such RNA structures and RNA sequences can greatly assist in RNA structure prediction.

The genetic algorithm has also been used to predict secondary structure; for aligning RNA sequences, taking into account both sequence and secondary structure and including pseudoknots; and for simulation of RNA-folding pathways. The program FOLDALIGN uses a dynamic programming algorithm to align RNAs based on sequence and secondary structure and locates the most significant motifs. Chan et al. (1991) have described another algorithm for the same purpose, and Chetouani et al. (1997) have developed ESSA, a method for viewing and analyzing RNA secondary structure.

Methods

Self-complementary Regions in RNA Sequences Predict Secondary Structure

One of the simplest types of analyses that can be performed to find stretches of sequence in RNA that are self-complementary is a dot matrix sequence comparison for self-complementary regions. For single-stranded RNA molecules, these repeats represent regions that can potentially self-hybridize to form RNA double strands. All types of RNA secondary structure analysis begin by the identification of these regions, and, once identified, the compatible regions may be used to predict a minimum free-energy structure. A more advanced type of dot matrix can be used to show the most energetic parts of the molecule.

Self-complementary regions in RNA may be found by performing a dot matrix analysis with the sequence to be analyzed listed in both the horizontal and vertical axes. In one method for finding such regions, the sequence is listed in the 5′→3′ direction across the top of the page and the sequence of the complementary strand is listed down the side of the page, also in the 5′→3′ direction. The matrix is then scored for identities. Self-complementary regions appear as rows of dots going from upper left to lower right. For RNA, these regions represent sequences that can potentially form A/U and G/C base pairs. G/U base pairs will not usually be included in this simple type of analysis. As with matching DNA sequences, there are many random

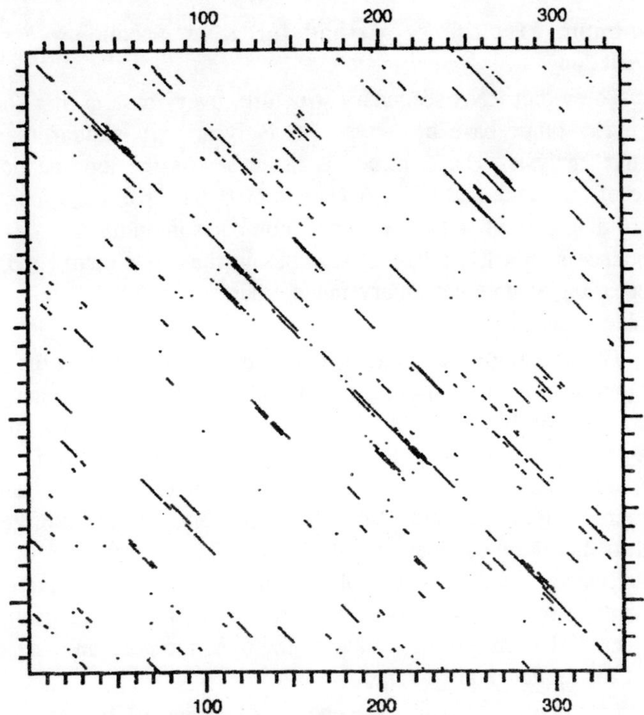

Fig. 6.4. Dot matrix analysis of the potato tuber spindle viroid for RNA secondary structure using the MATRIX function of DNA.

matches between the four bases in RNA, and the diagonals are difficult to visualize. A long window and a requirement for a large number of matches within this window are used to filter out these random matches.

An analysis of the potato spindle tuber viroid is shown, using a window of 15 and a required match of 11. Note the appearance of a diagonal running from the center of the matrix to the upper left, and a mirror image of this diagonal running to the lower right. The presence of this diagonal indicates the occurrence of a large self-complementary sequence such that the entire molecule can potentially fold into a hairpin structure. An alternative dot matrix method for finding RNA secondary structure is to list the given RNA sequence across the top of the page and also down the side of the page and then to score matches of complementary bases (G/C, A/U, and G /U). Diagonals indicating complementary regions will go from upper right to lower left in this type of matrix. This is the kind of matrix used to produce an energy matrix.

Minimum Free-energy Method for RNA Secondary Structure Prediction

To predict RNA secondary structure, every base is first compared to every other base by a type of analysis very similar to the dot matrix analysis. The sequence is listed across the top and down the side of the page, and G/C, A/U, and G/U base pairs are scored. Just as a diagonal in a two-sequence comparison indicates a range of sequence similarity, a row of matches in the RNA matrix indicates a succession of complementary nucleotides that can potentially form a double-stranded region.

The energy of each predicted structure is estimated by the nearest-neighbour rule by summing the negative base-stacking energies for each pair of bases in double-stranded regions and by adding the estimated positive energies of destabilizing regions such as loops at the end of hairpins, bulges within hairpins, internal bulges, and other unpaired regions. To evaluate all the different possible configurations and to find the most energetically favourable, several types of scoring matrices are used. The complementary regions are evaluated by a dynamic programming algorithm to predict the most energetically stable molecule. The method is similar to the dynamic programming method used for sequence alignment.

To calculate the stacking energy of a row of base pairs in the molecule, the stacking energies similar to those shown in Table 6.1 are used. An illustrative example for evaluation of energy in a double-stranded region is shown in Figure 6.5. The sequence is listed down the side of the matrix, and a portion of the same sequence is also listed across the top of the matrix; matching base pairs have been identified within the matrix. The object is to find a diagonal row of matches that goes from upper right to lower left, and such a row is shown in the example. In Figure 6.5, a match of four complementary bases in a row produces a molecule of free energy —6.4 kcal/mole.

In general, each matrix value is obtained by considering the minimum energy values obtained by all previous complementary pairs decreased by the stacking energy of any additional complementary base pairs or increased by the destabilizing energy associated with noncomplementary bases. The increase depends on the type and length of loop that is introduced by the noncomplementary base pair, whether internal loop, bulge loop, or hairpin loop. This comparison of all possible matches and energy values is continued until all nucleotides have been compared.

A. Base comparisons

5'	A	C	G	U	3'
A					
C					
G					
U					
–					
–					
G		C/G		U/G	
C					
G				U/G	
U	A/U	C/U	G/U		
3'					

B. Free energy calculations

5'	A	C	G	U	3'
A					
C					
G					
U					
–					
–					
G				−6.4	
C			−5.2		
G		−1.8			
U					
3'					

Fig. 6.5. Evaluation of secondary structure in RNA sequence.

Suboptimal Structure Predictions by MFOLD and the Use of Energy Plots

Originally, the FOLD program of M. Zuker predicted only one structure having the minimum free energy. However, changes in a single nucleotide can result in drastic changes in the predicted structure.

Table 6.1. Predicted free-energy values (kcal/mole at 37°C) for base pairs and other features of predicted RNA secondary structures

A. Stacking energies for base pairs

	A/U	*C/G*	*G/C*	*U/A*	*G/U*	*U/G*
A/U	-0.9	-1.8	-2.3	-1.1	-1.1	-0.8
C/G	-1.7	-2.9	-3.4	-2.3	-2.1	-1.4
G/C	-2.1	-2.0	-2.9	-1.8	-1.9	-1.2
U/A	-0.9	-1.7	-2.1	-0.9	-1.0	-0.5
G/U	-0.5	-1.2	-1.4	-0.8	-0.4	-0.2
U/G	-1.0	-1.9	-2.1	-1.1	-1.5	-0.4

B. Destabilizing energies for loops

Number of bases	*1*	*5*	*10*	*20*	*30*
Internal	–	5.3	6.6	7.0	7.4
Bulge	3.9	4.8	5.5	6.3	6.7
Hairpin	–	4.4	5.3	6.1	6.5

A later version, called MFOLD, has improved prediction of non-base-paired interactions and predicts several structures having energies close to the minimum free energy. These predictions accurately reflect structures of related RNA molecules derived from comparative sequence analysis. To find these suboptimal structures, the dynamic programming method was modified to evaluate parts of a new scoring matrix in which the sequence is represented in two tandem copies on both the vertical and horizontal axes. The regions from $i = 1$ to n and $j = 1$ to n are used to calculate an energy $V(i,j)$ for the best structure that includes an i,j base pair and is called the included region. A second region, the excluded region, is used to calculate the energy of the best structure that includes i,j but is not derived from the structure at $i+1$, $j-1$. After certain corrections are made, the difference between the included and excluded values is the most energetic structure that includes the base pair i,j. All complementary base pairs can be sampled in this fashion to determine which are present in a suboptimal structure that is within a certain range of the optimal one.

An energy dot plot is produced showing the locations of alternative base pairs that produce the most stable or suboptimally stable structures. The program may be instructed to find structures within a certain percentage of the minimum free energy. Parameter d provides a measure of similarity between two structures. When MFOLD is established on a suitable local host machine, the window is interactive, and clicking a part of the display will lead to program output of the corresponding structure. The dot plot may be filtered so that only suboptimal regions with helices of a certain minimal length are shown.

Other Algorithms for Suboptimal Folding of RNA Molecules

A limitation of the Zuker method and other methods for computing suboptimal RNA structures is that they do not compute all the structures within a given energy range of the minimum tree-energy structure. For example, no alternative structures are produced that have the absence of base pairs in the best structure, and, if two substructures are joined by a stretch of unpaired bases, no structures are produced that are suboptimal for both structures. These factors limit the number of alternative structures predicted compared to known variations based on sequence variations in tRNAs.

These limitations have been largely overcome by using an algorithm originally described by Waterman and Byers (1985) for finding sequence alignments within a certain range of the optimal one by modifications of the trace-back procedure used in dynamic programming. This method efficiently calculates a large number of alternative structures, up to a very large number, within a given energy range of the minimum free-energy structure. The method has been used to demonstrate that natural tRNA sequences can form many alternative structures which are close to the minimum free-energy structure and that base modification plays a major role in this energetic stability. The method may also be used to assess the thermodynamic stability of RNA structures given expected changes in energies associated with base pairs and loops as a function of temperature.

Prediction of Most Probable RNA Secondary Structure

In the above types of analyses, the energy associated with predicted double-stranded regions in RNA is used to produce a secondary structure. Stabilizing energies associated with base-paired regions and destabilizing energies associated with loops are summed to produce the most stable structure or suboptimal RNA secondary structure. A different way of predicting the structures is to consider the probability that each base-paired region will form based on principles of thermodynamics and statistical mechanics. The probability of forming a region with free energy ΔG is expressed by the Boltzmann distribution, which states that the likelihood of finding a structure with free energy $-\Delta G$ is proportional to [exp $(-\Delta G/kT)$] where k is the Boltzmann gas constant and T is the absolute temperature. Note that the more stable a structure, the lower the value of ΔG. Since ΔG is a negative number, the value of exp(-ΔG/kT) increases for more stable structures and also grows exponentially with a decrease in energy. The probability of these regions forming increases in the same manner. Conversely, the effect of destabilizing loops that have a positive ΔG is to decrease the probability of formation. By using these probability calculations and a dynamic programming method similar to that used in MFOLD, it is possible to predict the most probable RNA secondary structures and to assess the probability of the base pairs that contribute energetic stability to this structure.

For a set of possible structural states, the likelihood of each may be calculated using this formula, and the sum of these likelihoods provides a partition function that can be used to normalize each individual likelihood, providing a probability that each will occur. Thus, probability of structure A of energy $-\Delta G_a$ is [exp $(-\Delta G_a/kT)$] divided by the partition function Q, where $Q = \Sigma_s$, [exp $(-\Delta G_s/kT)$], the sum of probabilities of all possible structures, s. This kind of analysis allows one to calculate the probability of a certain base pair forming.

The key to this analysis is the calculation of the partition function Q. A dynamic programming algorithm for calculating this function exactly for RNA secondary structure has been developed. The algorithm is very similar to that used for computing an optimal folding by MFOLD. Complexity similarly increases as the cube of the sequence length, and the energy values used for base pairs and loops are also the same except that structures with very large interior loops are ignored. Just as the minimum free-energy value is given at $W(1,n)$ in the Zuker MFOLD algorithm, the value of the partition function is given at matrix position $Q(1,n)$ in the corresponding partition matrix.

As indicated above, the partition function is calculated as the sum of the probabilities of each possible secondary structure. Because there are a very large possible number of structures, the calculation is simplified by calculating an auxiliary function, $Q^b(i,j)$, which is the sum of the probabilities of all structures that include the base pair i,j. The partition function $Q(i,j)$ includes both these structures and the additional ones where i is not paired with j. An example illustrating the difference between the minimum free energy and the partition function methods should be instructive. Suppose that the bases at positions $i+1$, $j-1$ and i,j can both form base pairs. They then form a stack of two base pairs. In the minimum free-energy method, the energy of the i,j pair stacked on the $i+1$, $j-1$ pair will be added to $V(i+1, j-1)$ to give $V(i,j)$, where V is a scoring matrix that keeps track of the best structure that includes an i,j base pair. In contrast, the value for $Q^b(i,j)$ will be calculated by multiplying the matrix value $Q^b(i+1, j-1)$ by the probability of the base pair i,j given by the Boltzmann probability [exp $(-\Delta G/kT)$], where ΔG is the negative stacking energy of the i,j base pair on the $i+1$, $j-1$ base pair, and will be a large number reflecting the probability given the stability of the base-paired region.

For a hairpin structure with a row of successive base pairs, the probability will be the product of the Boltzmann factors associated with the stacked pair, giving a high number for the relative likelihood of formation. The procedure followed by the partition function algorithm is to calculate $Q^b(i,j)$ and $Q(i,j)$ iteratively in scoring matrix until $Q(1,n)$ is reached. This matrix position contains the value of the full partition function Q.

Both the partition function and the probabilities of all base pairs are computed by this algorithm, and the most probable structural model is thereby found. Information about intermediate structures, base-pair opening and slippage, and the temperature dependence of the partition function may also be determined. The latter calculation provides information about the melting behaviour of the secondary structure.

A suite of RNA-folding programs available from the Vienna RNA secondary structure prediction Web site uses this methodology to predict the most probable and alternative RNA secondary structures. The probability of forming each base pair is shown in a dot matrix display in which the dots are squares of increasing size reflecting the probability of the base pair formed by the bases in the horizontal and vertical positions of the matrix. Secondary structure prediction is done by two kinds of dynamic programming algorithms: the minimum free-energy algorithm of Zuker and Stiegler (1981) and the partition function algorithm of McCaskill (1990).

Using Sequence Covariation to Predict Structure

The second major method that has been used to make RNA secondary structure predictions and also tertiary structure analyses is RNA sequence covariation analysis. This method examines sequences of the same RNA molecules from different species for positions that vary together in a manner that would allow them to produce a base pair in all of the molecules. The idea is quite simple. On the one hand, for double-stranded regions in RNA molecules, sequence changes that take place in evolution should maintain the base-pairing. On the other hand, sequence changes in loops and single-stranded regions should not have such a constraint. The method of analysis is to look for sequence positions at which covariation maintains the base-pairing properties. The justification for this method is that these types of joint substitutions or covariations actually are found to occur during evolution of such genes. When one position corresponding to a base pair is changed, another position corresponding to the base-pairing partner will also change. For example, if two positions G and C form a base pair, then sequences that have C and G reversed, or A and T or T and A at the corresponding positions, would also be considered reasonable matches. Sequence covariability has been used to improve thermodynamic structure prediction.

One method of covariation analysis also examines which phylogenetic groups exhibit change at a given position. For each

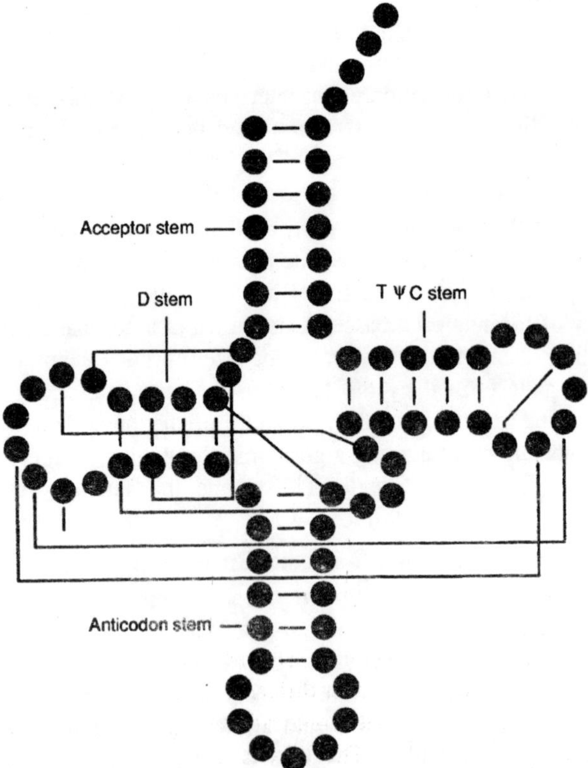

Fig. 6.6. Diagram of tRNA structure illustrating base-base interactions revealed by a covariance analysis.

position, the base that generally predominates in one particular part of the tree is determined. These methods have required manual examination of sequences and structures for covariation, but automatic methods have also been devised and demonstrated to produce reliable predictions.

A formal covariance model has been devised by Eddy and Durbin (1994). Although very accurate when used for identifying tRNA genes, the algorithm is extremely slow and unsuitable for searching through large genomes. Instead, the method has been used to screen through putative tRNA genes previously identified by faster methods. The difficulty that is faced in modeling RNA molecules is to identify the potential base pairs in a set of related RNA molecules based on covariation at two sites. Recall the hidden Markov model is used for capturing the types of variations observed in a sequence profile, including matches, mismatches, insertions, and deletions. This type of model assumes each sequence can be predicted by a series of states in the model, one after the other, as in a series of independent events in a Markov chain. The hidden Markov model does not analyze joint variations at sequence positions such as occur in RNA molecules. The model that is used for analyzing RNA secondary structure (but not tertiary structure) is an ordered tree model.

The above assumes that we know which bases are paired in a model of RNA secondary structure, whereas the goal is to build a model that discovers this information. The task is achieved by constructing a more general model, training the model with a set of sequences, and then having the model reveal the most likely base-paired regions. The approach is similar to training a hidden Markov model for proteins to recognize a family of protein sequences, thereby producing the most probable multiple sequence alignment. In the case of RNA secondary structure, a tree model is trained by the sequences, and the model may then be used to predict the most probable secondary structure. In addition, the model may also be used to search a database for sequences that produced a high score when aligned to the model. These sequences are likely to encode a similar type of RNA molecule such as tRNA or 5S RNA. Each model is derived by training a more general tree model with the sequences.

The general tree model needs to represent the types of variations that are found in aligning a series of related sequences, such as insertions, deletions, and mismatches. To allow for such variations, each node in the tree is replaced by a set of states that correspond to all of the possible sequence variations that might be encountered at that position. The mutual information content of all sequence positions is used in designing the model, and the expectation maximization method is used to optimize the parameters of the model. A dynamic programming method is used to find a model that maximizes the amount of covariation. The structure of the model may subsequently be altered during training. Once a covariance model suitable for an RNA molecule has been established, the model is trained by the sequences. The methodology is similar to that of hidden Markov models. Basically, the model is initialized by giving starting values to the base and dinucleotide frequencies in each MATCH and INS state and to the transition probabilities. All possible paths through the model are found for each sequence in the training set. The frequencies and transition probabilities are modified each time a particular path in the model is used. The base pairs are found from MATP, which gives probabilities to the 16 possible dinucleotides.

Once the model has been trained, the most probable path for each sequence provides a consensus structural alignment of the sequences. A dynamic programming algorithm is used that matches subsequence alignments to the nodes of the covariance model. The result is a log odds score of the sequence matching the covariance model. A similar method may be used to find sequences in a genomic database with high matching scores to the covariance model. The method was used to predict the structural alignment of representative sets of tRNA sequences, and it provided alignments that closely matched actual structural alignments based on other methods. The software for the COVELS program is available by request from the authors.

Stochastic Context-free Grammars for Modeling RNA Secondary Structure

In the above section, we discussed the need to have models for RNA secondary structure that reflect the interaction among base pairs. Simpler models of sequence variation treat sequences as simple strings of characters without such interactions and are therefore not suitable for RNA. A general theory for modeling strings of symbols, such as bases in DNA sequences, has been developed by linguists. There is a hierarchy of these so-called transformational grammars that deal with situations of increasing complexity. The application of the grammars to sequence analysis has been extensively discussed elsewhere. The context-free grammar is suitable for finding groups of symbols in different parts of the input sequence that thus are not in the same context. Complementary regions in sequences, such as those in RNA that will form secondary structures, are an example of such context-free sequences. Stochastic context-free grammars (SCFG) introduce uncertainty into the definition of such regions, allowing them to use alternative symbols as found in the evolution of RNA molecules. Thus, SCFGs can help define both the types of base interactions in specific classes of RNA molecules and the sequence variations at those positions. SCFGs have been used to model tRNA secondary structure. Although SCFGs are computationally complex, they are likely to play an important future role in identifying specific types of RNA molecules.

The application of SCFGs to RNA secondary structure analysis is very similar in form to the probabilistic covariance models. For RNA, the symbols of the alphabet are A, C, G, and U. The context-free grammar establishes a set of rules called productions for generating the sequence from the alphabet, in this case an RNA molecule with sections that can base-pair and others that cannot base-pair. In addition to the sequence symbols (named terminal symbols because they end up in the sequence), another set of symbols (nonterminal symbols) designated S_0, S_1, S_2 ..., determines intermediate production stages. The initial symbol is S_0 by convention. The next terminal symbol S_1 is produced by modifying S_0 in some fashion by productions indicated by an arrow. For example, the productions $S_0 \rightarrow S_1$, $S_1 \rightarrow C\ S_2\ G$ generate the sequence C S_2 G where S_2 has to be defined further by additional productions. The example shows a set of productions for generating the sequence CAUCAGGGAAGALJCUGUUG and also the secondary structure of this molecule. The productions chosen describe both features.

In this example of a context-free grammar, only one sequence is produced at each production level. In a SCFG, each production of a nonterminal symbol has an associated probability for giving rise to the resulting product, and there are a set of productions, each giving a different result. For example, the production $S_1 \rightarrow C\ S_2\ G$ could also be represented by 15 other base-pair combinations, and each of these has a corresponding probability. Thus, each production can be considered to be represented by a probability distribution over the possible outcomes. Note the identity of the SCFG representation of the predicted structure to that shown for the tree representation of the covariance model. The use of SCFGs in RNA secondary structure production analysis is in fact very similar to that of the covariance model, with the grammatical productions resembling the nodes in the ordered binary tree. As with hidden Markov models, the probability distribution of

A. Productions

$$
\begin{array}{ll}
P = \{ S_0 \rightarrow S_1, & S_7 \rightarrow G\,S_8, \\
S_1 \rightarrow C\,S_2\,G, & S_8 \rightarrow G, \\
S_2 \rightarrow A\,S_3\,U, & S_9 \rightarrow A\,S_{10}\,U, \\
S_3 \rightarrow S_4\ S_9, & S_{10} \rightarrow G\,S_{11}\,C, \\
S_4 \rightarrow U\,S_5\,A, & S_{11} \rightarrow A\,S_{12}\,U, \\
S_5 \rightarrow C\,S_6\,G, & S_{12} \rightarrow U\,S_{13}, \\
S_6 \rightarrow A\,S_7, & S_{13} \rightarrow C \quad \}
\end{array}
$$

B. Derivation

$$
\begin{aligned}
S_0 &\rightarrow S_1 \rightarrow CS_2G \rightarrow CAS_3UG \rightarrow CAS_4S_9UG \\
&\rightarrow CAUS_5AS_9UG \rightarrow CAUCS_6GAS_9UG \\
&\rightarrow CAUCAS_7GAS_9UG \rightarrow CAUCAGS_8GAS_9UG \\
&\rightarrow CAUCAGGGAS_9UG \rightarrow CAUCAGGGAAS_{10}UUG \\
&\rightarrow CAUCAGGGAAGS_{11}CUUG \\
&\rightarrow CAUCAGGGAAGAS_{12}UCUUG \\
&\rightarrow CAUCAGGGAAGAUS_{13}UCUUG \\
&\rightarrow CAUCAGGGAAGAUCUCUUG.
\end{aligned}
$$

C. Parse tree

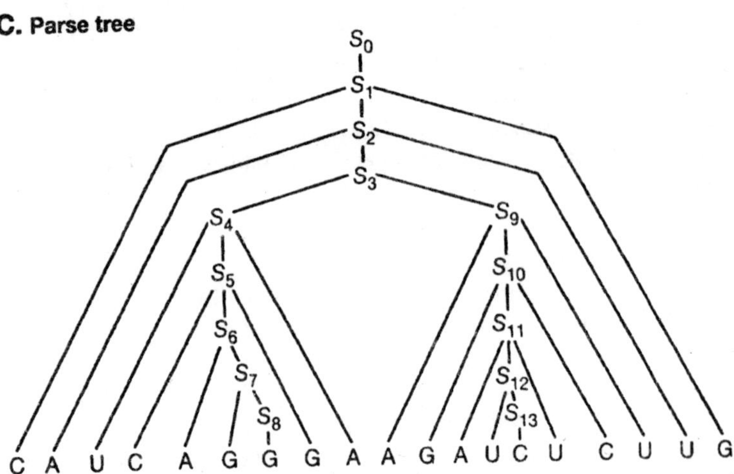

Fig. 6.7. A set of transformation rules for generating an RNA sequence and the secondary structure of the sequence from the RNA alphabet.

each production must be derived by training with known sequences. The algorithms used for training the SCFG and for aligning a sequence with the SCFG are somewhat different from those used with hidden Markov models, and the time and memory requirements are greater.

Searching Genomes for RNA-specifying Genes

One goal in RNA research has been to design methods to identify sequences in genomes that encode small RNA molecules. Larger, highly conserved molecules can simply be identified based on their sequence similarity with already-known sequences. For smaller sequences with more sequence variation, this method does not work. A number of methods for finding small RNA genes have been described and are available on the Web. A major problem with these methods in searches of large genomes is that a small false-positive rate becomes quite unacceptable because there are so many false positives to check out.

One of the first methods used to find tRNA genes was to search for sequences that are self-complementary and can fold into a hairpin like the three found in tRNAs. Fichant and Burks (1991) described a program, tRNAscan, that searches a genomic sequence with a sliding window searching simultaneously for matches to a set of invariant bases and conserved self-complementary regions in tRNAs with an accuracy of 97.5%. Pavesi et al. (1994) derived a method for finding the RNA polymerase III transcriptional control regions of tRNA genes using a scoring matrix derived from known control regions that is also very accurate. Finally, Lowe and Eddy (1997) have devised a search algorithm tRNAscan-SE that uses a combination of three methods to find tRNA genes in genomic sequences—tRNAscan, the Pavesi algorithm, and the COVELS program based on sequence covariance analysis. This method is reportedly 99-100% accurate with an extremely low rate of false positives.

The probabilistic model was used to identify small nucleolar (sno) RNAs in the yeast genome that methylate ribosomal RNA. The model is not used to search genomic sequences directly. Instead, a list of candidate sequences is first found by searching for patterns that match the sequences in the model. The probability model was a hybrid combination of HMMs and SCFGs trained on snoRNAs. These RNAs vary sufficiently in sequence and structure that they are not found by straightforward similarity searches. The RNAs found were shown to be snoRNAs by insertional mutagenesis.

Application of RNA Structure Modeling

In summary, methods for predicting the structure of RNA molecules include (1) an analysis of all possible combinations of potential double-stranded regions by energy minimization methods and (2) identification of base covariation that maintains secondary and tertiary structure of an RNA molecule during evolution. Energy minimization methods have been so well refined that a series of energetically feasible models and the most thermodynamically probable structural models may be computed. Covariation analysis by C. Woese led to his building of detailed structural models for rRNAs. By examining the evolutionary variation in these structures, he was able to predict three domains of life—the Bacteria, the Eukarya, and a newly identified Archaea. Although a large amount of horizontal transfer among evolutionary lineages of other genes has added a great deal of noise to the evolutionary signal, the rRNA-based prediction is supported by other types of genomic analyses. In addition to these uses of rRNA structural analysis, excellent probabilistic models of two small RNA molecules, tRNA and snoRNA, have been built, and these models may be used to search reliably through genomic sequences for genes that encode these RNA molecules. The successful analysis of these types of RNA molecules should be readily extensible to other classes of RNA molecules.

7

CLUSTERING AND TREES

A common problem in biology is to partition a set of experimental data into groups (clusters) in such a way that the data points within the same cluster are highly similar while data points in different clusters are very different. This problem is far from simple, and this chapter covers several algorithms that perform different types of clustering. There is no simple recipe for choosing one particular approach over another for a particular clustering problem, just as there is no universal notion of what constitutes a "good cluster." Nonetheless, these algorithms can yield significant insight into data and allow one, for example, to identify clusters of genes with similar functions even when it is not clear what particular role these genes play. We conclude this chapter with studies of evolutionary tree reconstruction, which is closely related to clustering.

GENE EXPRESSION ANALYSIS

Sequence comparison often helps to discover the function of a newly sequenced gene by finding similarities between the new gene and previously sequenced genes with known functions. However, for many genes, the sequence similarity of genes in a functional family is so weak that one cannot reliably derive the function of the newly sequenced gene based on sequence alone. Moreover, genes with the same function sometimes have no sequence similarity at all. As a result, the functions of more than 40% of the genes in sequenced genomes are still unknown.

In the last decade, a new approach to analyzing gene functions has emerged. DNA arrays allow one to analyze the *expression levels* (amount of mRNA produced in the cell) of many genes under many time points and conditions and to reveal which genes are switched on and switched off in the cell. The outcome of this type of study is an $n \times m$ *expression matrix* I, with the n rows corresponding to genes, and the m columns corresponding to different time points and different conditions. The expression matrix I represents *intensities* of hybridization signals as provided by a DNA array, in reality, expression matrices usually represent transformed and normalized intensities rather than the raw intensities obtained as a result of a DNA array experiment, but we will not discuss this transformation.

The element $I_{i,j}$ of the expression matrix represents the expression level of gene i in experiment j; the entire ith row of the expression matrix is called the *expression pattern* of gene i. One can look for pairs of genes in an expression matrix with similar expression patterns, which would be manifested as two similar rows. Therefore, if the expression patterns of two genes are similar, there is a good chance that these genes are somehow related, that is, they either perform similar functions or are involved in the same biological process. Accordingly, if the expression pattern of a newly sequenced gene is similar to the expression pattern of a gene with known function, a biologist may have reason to suspect that these genes perform similar or related functions. Another important application of expression analysis is in the deciphering of regulatory pathways; similar expression patterns usually imply coregulation. However, expression analysis should be done with caution since DNA arrays typically produce noisy data with high error rates.

Clustering algorithms group genes with similar expression patterns into *clusters* with the hope that these clusters correspond to groups of functionally related genes. To cluster the expression data, the $n \times m$ expression matrix is often transformed into an $n \times n$ *distance matrix* $d = (d_{i,j})$ where $d_{i,j}$ reflects how similar the expression patterns of genes i and j are. The goal of clustering is to group genes into clusters satisfying the following two conditions:

1. *Homogeneity*. Genes (rather, their expression patterns) within a cluster should be highly similar to each other. That is, $d_{i,j}$ should be small if i and j belong to the same cluster.
2. *Separation*. Genes from different clusters should be very different. That is, $d_{i,j}$ should be large if i and j belong to different clusters.

A "good" clustering of data is one that adheres to these goals. While we hope that a better clustering of genes gives rise to a better grouping of genes on a functional level, the final analysis of resulting

Time	1 hr	2 hr	3 hr
g_1	10.0	8.0	10.0
g_2	10.0	0.0	9.0
g_3	4.0	8.5	3.0
g_4	9.5	0.5	8.5
g_5	4.5	8.5	2.5
g_6	10.5	9.0	12.0
g_7	5.0	8.5	11.0
g_8	2.7	8.7	2.0
g_9	9.7	2.0	9.0
g_{10}	10.2	1.0	9.2

(a) Intensity matrix, I

	g_1	g_2	g_3	g_4	g_5	g_6	g_7	g_8	g_9	g_{10}
g_1	0.0	8.1	9.2	7.7	9.3	2.3	5.1	10.2	6.1	7.0
g_2	8.1	0.0	12.0	0.9	12.0	9.5	10.1	12.8	2.0	1.0
g_3	9.2	12.0	0.0	11.2	0.7	11.1	8.1	1.1	10.5	11.5
g_4	7.7	0.9	11.2	0.0	11.2	9.2	9.5	12.0	1.6	1.1
g_5	9.3	12.0	0.7	11.2	0.0	11.2	8.5	1.0	10.6	11.6
g_6	2.3	9.5	11.1	9.2	11.2	0.0	5.6	12.1	7.7	8.5
g_7	5.1	10.1	8.1	9.5	8.5	5.6	0.0	9.1	8.3	9.3
g_8	10.2	12.8	1.1	12.0	1.0	12.1	9.1	0.0	11.4	12.4
g_9	6.1	2.0	10.5	1.6	10.6	7.7	8.3	11.4	0.0	1.1
g_{10}	7.0	1.0	11.5	1.1	11.6	8.5	9.3	12.4	1.1	0.0

(b) Distance matrix, d

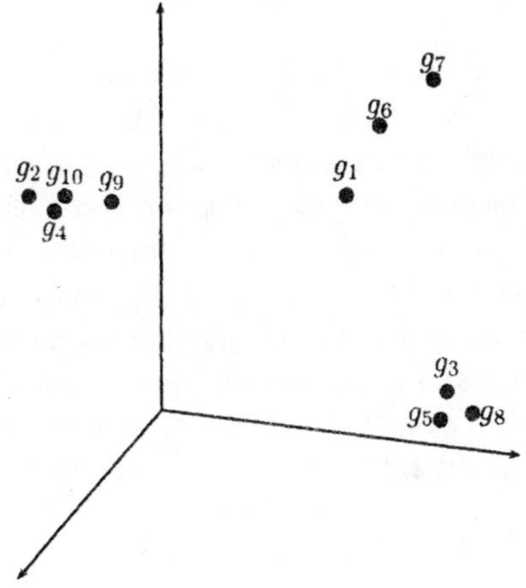

(c) Expression patterns as points in three-dimentsional space.

Fig. 7.1. An "expression" matrix of ten genes measured at three time points, and the corresponding distance matrix.

clusters is left to biologists. Different tissues express different genes, and there are typically over 10,000 genes expressed in any one tissue. Since there are about 100 different tissue types, and since expression levels are often measured over many time points, gene expression experiments can generate vast amounts of data which can be hard to interpret. Compounding these difficulties, expression levels of related genes may vary by several orders of magnitude, thus creating the problem of achieving accurate measurements over a large range of expression levels; genes with low expression levels may be related to genes with high expression levels.

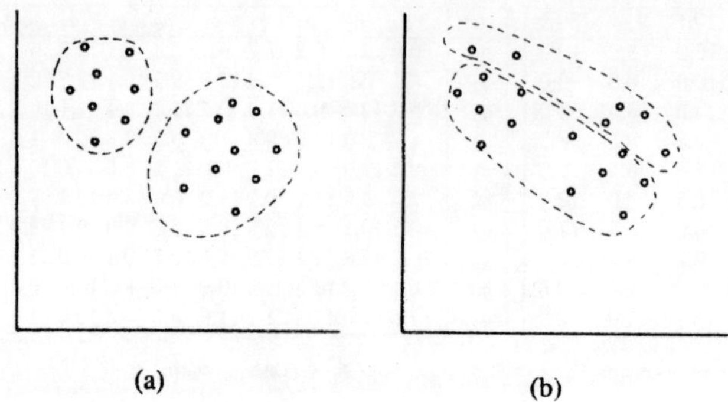

Fig. 7.2. Data can be grouped into clusters. Some clusters are better than others: the two clusters in (a) exhibit good homogeneity and separation, while the clusters in (b) do not.

Hierarchical Clustering

In many cases clusters have subclusters, these have subsubclusters, and so on. For example, mammals can be broken down into primates, carnivora, bats, marsupials, and many other orders. The order carnivora can be further broken down into cats, hyenas, bears, seals, and many others. Finally, cats can be broken into thirty seven species.

Hierarchical clustering is a technique that organizes elements into a tree, rather than forming an explicit partitioning of the elements into clusters. In this case, the genes are represented as the leaves of a tree. The edges of the trees are assigned *lengths* and the distances between leaves—that is, the length of the path in the tree that connects two leaves—correlate with entries in the distance matrix, Such trees are used in both the analysis of expression data and in studies of molecular evolution which we will discuss below.

This tree actually describes a family of different partitions into clusters, each with a different number of clusters (one for each value from 1 to n). You can see what these partitions by drawing a horizontal line through the tree. Each line crosses the tree at i points ($1 \leq i \leq k$) and correspond to i clusters.

The *hierarchical clustering* algorithm below takes an $n \times n$ distance matrix d as an input, and progressively generates n different partitions of the data as the tree it outputs. The largest partition has n single-element clusters, with every element forming its own cluster. The second-largest partition combines the two closest clusters from the largest partition, and thus has $n - 1$ clusters. In general, the ith partition combines the two closest clusters from the $(i - 1)$th partition and has $n - i + 1$ clusters.

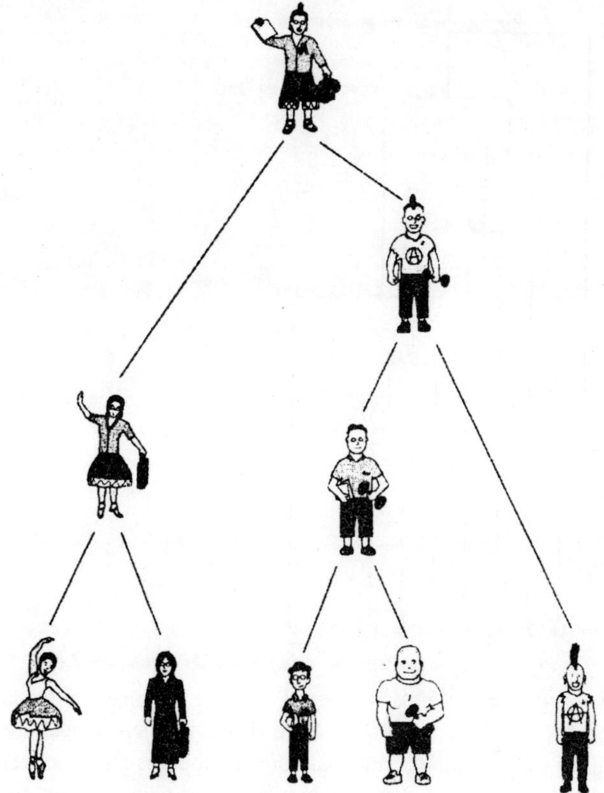

Fig. 7.3. A schematic of hierarchial clustering.

Hierarchical clustering (d, n)

```
1 Form n clusters, each with 1 element
2 Construct a graph T by assigning an isolated vertex to each cluster
3 while there is more than 1 cluster
4       Find the two closest clusters C1 and C2
5       Merge C1 and C2 into new cluster C with |C1| + |C2| elements
6       Compute distance from C to all other clusters
7       Add a new vertex C to T and connect to vertices C1 and C2
8       Remove rows and columns of d corresponding to C1 and C2
9       Add a row and column to d for the new cluster C
10 return T
```

Line 6 in the algorithm is (intentionally) left ambiguous; clustering algorithms vary in how they compute the distance between the newly

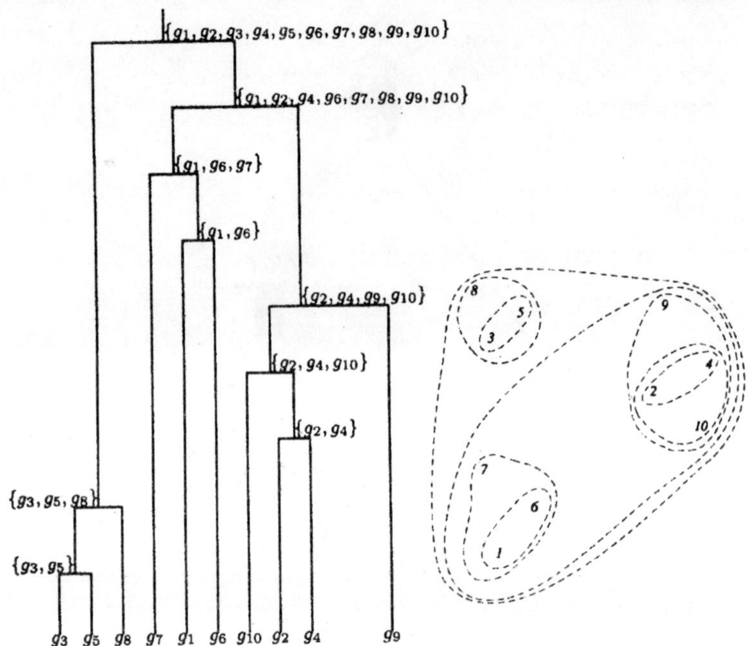

Fig. 7.4. A hierarchial clustering of the data in figure 7.1.

formed cluster and any other cluster. Different formulas for recomputing distances yield different answers from the same hierarchical clustering algorithm. For example, one can define the distance between two clusters as the smallest distance between any pair of their elements

$$d_{\min}(C^*, C) = \min_{x \in C^*, y \in C} d(x, y)$$

or the average distance between their elements

$$d_{avg}(C^*, C) = \frac{1}{|C^*||C|} \sum_{x \in C^*, y \in C} d(x, y).$$

Another distance function estimates distance based on the separation of C_1 and C_2 in *Hierarchical clustering*:

$$d(C^*, C) = \frac{d(C^*, C_1) + d(C^*, C_2) - d(C_1, C_2)}{2}$$

In one of the first expression analysis studies, Michael Eisen and colleagues used hierarchical clustering to analyze the expression profiles of 8600 genes over thirteen time points to find the genes responsible for the growth response of starved human cells. The *Hierarchical clustering* resulted in a tree consisting of five main subtrees and many smaller subtrees. The genes within these five clusters had similar functions, thus confirming that the resulting clusters are biologically sensible.

k-Means Clustering

One can view the n rows of the $n \times m$ expression matrix as a set of n points in m-dimensional space and partition them into k subsets, pretending that k—the number of clusters—is known in advance.

One of the most popular clustering methods for points in multidimensional spaces is called *k-Means clustering*. Given a set of n data points in m-dimensional space and an integer k, the problem is to determine a set of k points, or *centers*, in m-dimensional space that minimize the *squared error distortion* defined below. Given a data point υ and a set of k centers $X = \{x_1, \ldots x_k\}$, define the distance from υ to the centers X as the distance from υ to the closest point in X, that is, $d(\upsilon, X) = \min_{1 \leq i \leq k} d(\upsilon, x_i)$. We will assume for now that $d(\upsilon, x_i)$ is just the Euclidean distance in m dimensions. The squared error distortion for a set of n points $V = \{\upsilon_1, \ldots \upsilon_n\}$, and a set of k centers $X = \{x_1, \ldots x_k\}$, is defined as the mean squared distance from each data point to its nearest center:

$$d(V,X) = \frac{\sum_{i=1}^{n} d(\upsilon_i, X)^2}{n}$$

While the above formulation does not explicitly address *clustering* n points into k clusters, a clustering can be obtained by simply assigning each point to its closest center. Although the k-Means Clustering problem looks relatively simple, there are no efficient (polynomial) algorithms known for it. The *Lloyd* k-Means clustering algorithm is one of the most popular clustering heuristics that often generates good solutions in gene expression analysis. The Lloyd algorithm randomly selects an arbitrary partition of points into k clusters and tries to improve this partition by moving some points between clusters. In the beginning one can choose arbitrary k points as "cluster representatives." The algorithm iteratively performs the following two steps until either it converges or until the fluctuations become very small:

1. Assign each data point to the cluster C_i corresponding to the closest cluster representative x_i $(1 \leq i \leq k)$
2. After the assignments of all n data points, compute new cluster representatives according to the center of gravity of each cluster, that is, the new cluster representative is $\frac{\sum_{\upsilon \in C} \upsilon}{|C|}$ for every cluster C.

The Lloyd algorithm often converges to a local minimum of the squared error distortion function rather than the global minimum. Unfortunately, interesting objective functions other than the squared error distortion lead to similarly difficult problems. For example, finding a good clustering can be quite difficult if, instead of the squared error distortion $(\sum_{i=1}^{n} d(\upsilon_i, X)^2$, one tries to minimize $(\sum_{i=1}^{n} d(\upsilon_i, X)$ (*k-Median problem*) or $\max_{1 \le i \le n} d(\upsilon_i, X)$ (*k-Center problem*). We remark that all of these definitions of clustering cost emphasize the homogeneity condition and more or less ignore the other important goal of clustering, the separation condition. Moreover, in some unlucky instances of the *k*-Means Clustering problem, the algorithm may converge to a local minimum that is arbitrarily bad compared to an optimal solution.

While the Lloyd algorithm is very fast, it can significantly rearrange every cluster in every iteration. A more conservative approach is to move only one element between clusters in each iteration. We assume that every partition P of the n-element set into k clusters has an associated *clustering cost*, denoted *cost*(P), that measures the quality of the partition P: the smaller the clustering cost of a partition, the better that clustering is. The squared error distortion is one particular choice of *cost*(P) and assumes that each center point is the center of gravity of its cluster. The pseudocode below implicitly assumes that *cost*(P) can be efficiently computed based either on the distance matrix or on the expression matrix. Given a partition P, a cluster C within this partition, and an element i outside C, $P_{i \to C}$ denotes the partition obtained from P by moving the element i from its cluster to C. This move improves the clustering cost only if $\Delta(i \to C) = cost(P) - cost\ P_{i \to C}) > 0$, and the *ProgressiveGreedyK-means* algorithm searches for the "best" move in each step (i.e., a move that maximizes $\Delta(i \to C)$ for all C and for all $i \notin C$).

ProgressiveGreedyK-means (k)

```
Select an arbitrary partition P into k clusters.
while forever
    bestChange ← 0
    for every cluster C
        for every element i ∉ C
            if moving i to cluster C reduces the clustering cost
                if Δ(i → C) > bestChange
                    bestChange ← Δ(i → C)
                    i* ← i
```

```
10          C* ← C
11     if bestChange > 0
12        change partition P by moving i* to C*
13     else
14        return P
```

Even though line 2 makes an impression that this algorithm may loop endlessly, the return statement on line 14 saves us from an infinitely long wait. We stop iterating when no move allows for an improvement in the score; this eventually has to happen.

CLUSTERING AND CORRUPTED CLIQUES

A complete graph, written K_n, is an (undirected) graph on n vertices with every two vertices connected by an edge. A *clique graph* is a graph in which every connected component is a complete graph. Clique graph consisting of three connected components, K_3, K_5, and K_6. Every partition of n elements into k clusters can be represented by a clique graph on n vertices with k cliques. A subset of vertices $V' \subset V$ in a graph $G(V, E)$ forms a *complete subgraph* if the *induced subgraph* on these vertices is complete, that is, every two vertices υ and w in V' are connected by an edge in the graph. For example,

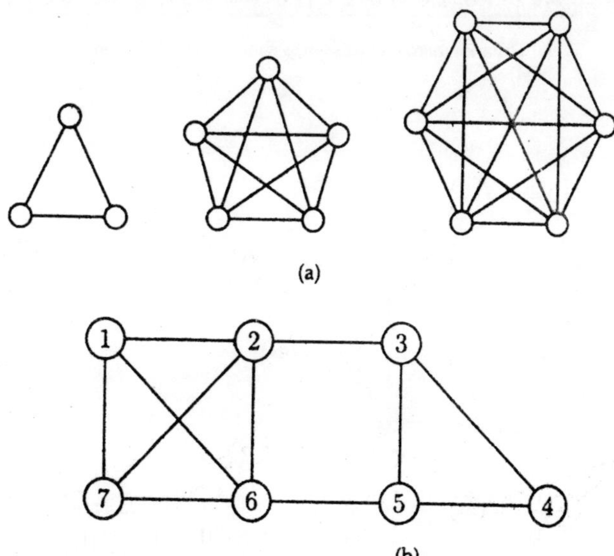

Fig. 7.5. (a) A clique graph consisting of the three connected components K_3, K_5, and K_6; (b) A graph with 7 vertices that has 4 cliques formed by vertices (1, 2, 6, 7), {2, 3}, {5, 6}, and {3, 4, 5}.

vertices 1, 6, and 7 form a complete subgraph of the graph. A *clique* in the graph is a maximal complete subgraph, that is, a complete subgraph that is not contained inside any other complete subgraph. For example, vertices 1, 6, and 7 form a complete subgraph but do not form a clique, but vertices 1, 2,6, and 7 do.

In expression analysis studies, the distance matrix ($d_{i,j}$) is often further transformed into a *distance graph* $G = G(\theta)$, where the vertices are genes and there is an edge between genes i and j if and only if the distance between them is below the threshold θ, that is, if $d_{i,j} < \theta$. A clustering of genes that satisfies the homogeneity and separation

	g_1	g_2	g_3	g_4	g_5	g_6	g_7	g_8	g_9	g_{10}
g_1	0.0	8.1	9.2	7.7	9.3	**2.3**	**5.1**	10.2	**6.1**	**7.0**
g_2	8.1	0.0	12.0	**0.9**	12.0	9.5	10.1	12.8	**2.0**	**1.0**
g_3	9.2	12.0	0.0	11.2	**0.7**	11.1	8.1	**1.1**	10.5	11.5
g_4	7.7	**0.9**	11.2	0.0	11.2	9.2	9.5	12.0	**1.6**	1.1
g_5	9.3	12.0	**0.7**	11.2	0.0	11.2	8.5	**1.0**	10.6	11.6
g_6	**2.3**	9.5	11.1	9.2	11.2	0.0	**5.6**	12.1	7.7	8.5
g_7	**5.1**	10.1	8.1	9.5	8.5	**5.6**	0.0	9.1	8.3	9.3
g_8	10.2	12.8	**1.1**	12.0	**1.0**	12.1	9.1	0.0	11.4	12.4
g_9	**6.1**	**2.0**	10.5	**1.6**	10.6	7.7	8.3	11.4	0.0	**1.1**
g_{10}	**7.0**	**1.0**	11.5	**1.1**	11.6	8.5	9.3	12.4	**1.1**	0.0

(a) Distance matrix, **d** (distances shorter than 7 are shown in bold).

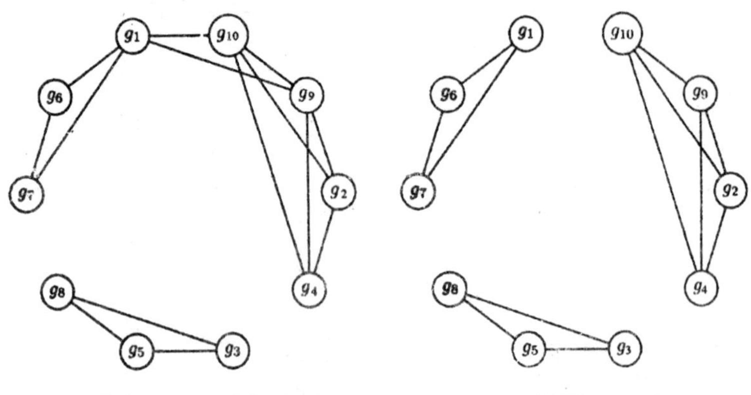

(b) Distance graph for $\theta = 7$. (c) Clique graph.

Fig. 7.6. The distance graph (b) for θ=7 is not quite a clique graph. However, it can be transformed into a clique graph (c) by removing edges (g_1, g_{10}) and (g_1, g_9).

principles for an appropriately chosen θ will correspond to a distance graph that is also a clique graph. However, errors in expression data, and the absence of a "universally good" threshold θ often results in distance graphs that do not quite form clique graphs. Some elements of the distance matrix may fall below the distance threshold for unrelated genes (adding edges between different clusters), while other elements of the distance matrix exceed the distance threshold for related genes (removing edges within clusters). Such erroneous edges corrupt the clique graph, raising the question of how to transform the distance graph into a clique graph using the smallest number of edge removals and edge additions.

It turns out that the Corrupted Cliques problem is *NP*-hard, so some heuristics have been proposed to approximately solve it. Below we describe the time-consuming PCC (*Parallel Classification with Cores*) algorithm, and the less theoretically sound, but practical, CAST (*Cluster Affinity Search Technique*) algorithm inspired by PCC.

Suppose we attempt to cluster a set of genes S, and suppose further that S' is a subset of S. If we are somehow magically given the correct clustering $\{C_1,\ldots,C_k\}$ of S', could we extend this clustering of S' into a clustering of the entire gene set S? Let $S \setminus S'$ be the set of unclustered genes, and $N(j, C_i)$ be the number of edges between gene $j \in S \setminus S'$ and genes from the cluster C_i in the distance graph. We define the *affinity* of gene j to cluster C_i as $\frac{N(j,C_i)}{|C_i|}$. A natural *maximal affinity* approach would be to put every unclustered gene j into the cluster C_i with the highest affinity to j, that is, the cluster that maximizes $\frac{N(j,C_i)}{|C_i|}$. In this way, the clustering of S' can be extended into clustering of the entire gene set S. In 1999, Amir Ben-Dor and colleagues developed the PCC clustering algorithm, which relies on the assumption that if S' is sufficiently large and the clustering of S' is correct, then the clustering of the entire gene set is likely to be correct.

The only problem is that the correct clustering of S' is unknown! The way around this is to generate *all* possible clusterings of S', extend them into a clustering of the entire gene set S, and select the resulting clustering with the best clustering score. As attractive as it may sound, this is not practical (unless S' is very small) since the number of possible partitions of a set S' into k clusters is $k^{|S'|}$. The PCC algorithm gets around this problem by making S' extremely small and generating *all* partitions of this set into k clusters. It then extends each of these $k^{|S'|}$ partitions into a partition of the entire n-element gene set by the two-stage procedure described below. The distance graph G guides these extensions based on the maximal affinity approach. The function *score(P)* is defined to be the number of edges necessary to add or remove to turn G into a clique graph according to the partition P. The PCC algorithm below clusters the set of elements S into k clusters according to the graph G by extending partitions of subsets of S using the maximal affinity approach:

```
PCC(G, k)
S ← set of vertices in the distance graph G
n ← number of elements in S
bestScore ← ∞
Randomly select a "very small" set S′ ⊂ S, where |S′| = log log n
Randomly select a "small" set S′′ ⊂ (S\S′), where |S′′| = log n.
for every partition P′ of S′ into k clusters
    Extend P′ into a partition P′′ of S′′
    Extend P′′ into a partition P of S
    if score(P) < bestScore
        bestScore ← score(P)
        bestPartition ← P
return bestPartition
```

The number of iterations that PCC requires is given by the number of partitions of set S', which is $k^{|S'|} = k^{\log\log n} = (\log n)^{\log_2 k} = (\log n)^C$. The amount of work done in each iteration is $O(n^2)$, resulting in a running time of $O(n^2(\log n)^C)$. Since this is too slow for most applications, a more practical heuristic called CAST is often used.

Define the distance between gene i and cluster C as the average distance between i and all genes in the cluster C: $d(i, C) = \frac{\sum_{j \in C} d(i,j)}{|C|}$. Given a threshold θ, a gene i is *close* to cluster C if $d(i, C) < \theta$ and *distant* otherwise. The CAST algorithm below clusters set S according to the distance graph G and the threshold θ. CAST iteratively builds the partition P of the set S by finding a cluster C such that no gene $i \notin C$ is close to C, and no gene $i \in C$ is distant from C. In the beginning of the routine, P is initialized to an empty set.

```
CAST(G, θ)
S ← set of vertices in the distance graph G
P ← ∅
```

```
while S ≠ ∅
    υ ← vertex of maximal degree in the distance graph G.
    C ← {υ}
    while there exists a close gene i ∉ C or distant gene i ∈ C
        Find the nearest close gene i ∉ C and add it to C
        Find the farthest distant gene i ∈ C and remove it from C
    Add cluster C to the partition P
    S ← S\C
    Remove vertices of cluster C from the distance graph G
return P
```

Although CAST is a heuristic with no performance guarantee it performs remarkably well with gene expression data.

Evolutionary Trees

In the past, biologists relied on morphological features, like beak shapes or the presence or absence of fins to construct evolutionary trees. Today biologists rely on DNA sequences for the reconstruction of evolutionary trees. This question is not as obvious as it may at first sound, since bears and raccoons diverged just 35 million years ago and they share many morphological features.

For over a hundred years biologists could not agree on whether the giant panda should be classified in the bear family or in the raccoon family. In 1870 an amateur naturalist and missionary, Pere Armand David, returned to Paris from China with the bones of the mysterious creature which he called simply "black and white bear." Biologists examined the bones and concluded that they more closely resembled the bones of a red panda than those of bears. Since red pandas were, beyond doubt, part of the raccoon family, giant pandas were also classified as raccoons (albeit big ones).

Although giant pandas look like bears, they have features that arc unusual for bears and typical of raccoons: they do not hibernate in the winter like other bears do, their male genitalia are tiny and backward-pointing (like raccoons' genitalia), and they do not roar like bears but bleat like raccoons. As a result, Edwin Colbert wrote in 1938:

> So the quest has stood for many years with the bear proponents and the raccoon adherents and the middle-of-the-road group advancing their several arguments with the clearest of logic, while in the meantime the giant panda lives serenely in the mountains of Szechuan with never a thought about the zoological controversies he is causing by just being himself.

The giant panda classification was finally resolved in 1985 by Steven O'Brien and colleagues who used DNA sequences and algorithms, rather than behavioural and anatomical features, to resolve the giant panda controversy. The final analysis demonstrated that DNA sequences provide an important source of information to test evolutionary hypotheses. O'Brien's study used about 500,000 nucleotides to construct the evolutionary tree of bears and raccoons.

Roughly at the same time that Steven O'Brien resolved the giant panda controversy, Rebecca Cann, Mark Stoneking and Allan Wilson constructed an evolutionary tree of humans and instantly created a new controversy. This tree led to the *Out of Africa* hypothesis, which claims that humans have a common ancestor who lived in Africa 200,000 years ago. This study turned the question of human origins into an algorithmic puzzle.

The tree was constructed from mitochondrial DNA (mtDNA) sequences of people of different races and nationalities. Wilson and his colleagues compared sequences of mtDNA from people representing African, Asian, Australian, Caucasian, and New Guinean ethnic groups and found 133 variants of mtDNA. Next, they constructed the evolutionary tree for these DNA sequences that showed a trunk splitting into two major branches. One branch consisted only of Africans, the other included some modern Africans and some people from everywhere else. They concluded that a population of Africans, the first modern humans, forms the trunk and the first branch of the tree while the second branch represents a subgroup that left Africa and later spread out to the rest of the world. All of the mtDNA, even samples from regions of the world far away from Africa, were strikingly similar. This suggested that our species is relatively young. But the African samples had the most mutations, thus implying that the African lineage is the oldest and that all modern humans trace their roots back to Africa. They further estimated that modern man emerged from Africa 200,000 years ago with racial differences arising only 50,000 years ago.

Shortly after Allan Wilson and colleagues constructed the human mtDNA evolutionary tree supporting the Out of Africa hypothesis, Alan Templeton constructed 100 distinct trees that were also consistent with data that provide evidence *against* the African origin hypothesis! This is a cautionary tale suggesting that one should proceed carefully when constructing large evolutionary trees' and below we describe some algorithms for evolutionary tree reconstruction.

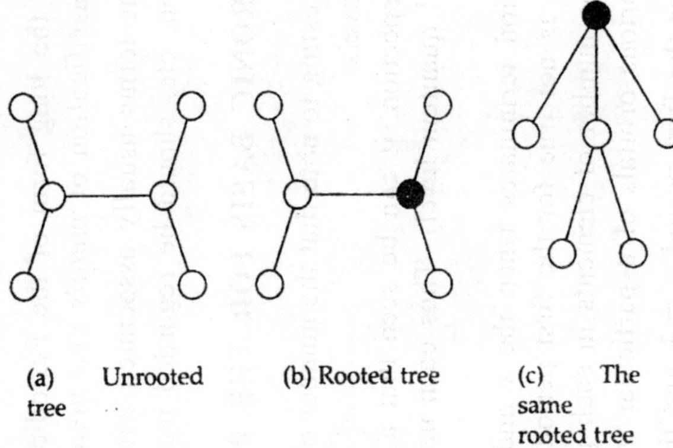

(a) Unrooted tree (b) Rooted tree (c) The same rooted tree

Fig. 7.7. The difference between unrooted (a) and rooted (b) trees.

Biologists use either *unrooted* or *rooted* evolutionary trees. In a rooted evolutionary tree, the root corresponds to the most ancient ancestor in the tree, and the path from the root to a leaf in the rooted tree is called an *evolutionary path*. Leaves of evolutionary trees correspond to the existing species while internal vertices correspond to hypothetical ancestral species. In the unrooted case, we do not make any assumption about the position of an evolutionary ancestor (root) in the tree. We also remark that rooted trees (defined formally as undirected graphs) can be viewed as directed graphs if one directs the edges of the rooted tree from the root to the leaves.

Biologists often work with *binary weighted* trees where every internal vertex has degree equal to 3 and every edge has an assigned positive weight (sometimes referred to as the *length*). The *weight* of an edge (υ, w) may reflect the number of mutations on the evolutionary path from υ to w or a time estimate for the evolution of species υ into species w. We sometimes assume the existence of a *molecular clock* that assigns a time $t(\upsilon)$ to every internal vertex υ in the tree and a length of $t(w) - t(\upsilon)$ to an edge (υ, w). Here, time corresponds to the "moment" when the species υ produced its descendants; every leaf species corresponds to time 0 and every internal vertex presumably corresponds to some negative time.

Distance-Based Tree Reconstruction

If we are given a weighted tree T with n leaves, we can compute the length of the path between any two leaves i and j, $d_{i,j}(T)$. Evolutionary biologists often face the opposite problem: they measure the $n \times n$ *distance matrix* $(D_{i,j})$, and then must search for a tree T that has n leaves and fits the data, that is, $d_{i,j}(T) = D_{i,j}$ for every two

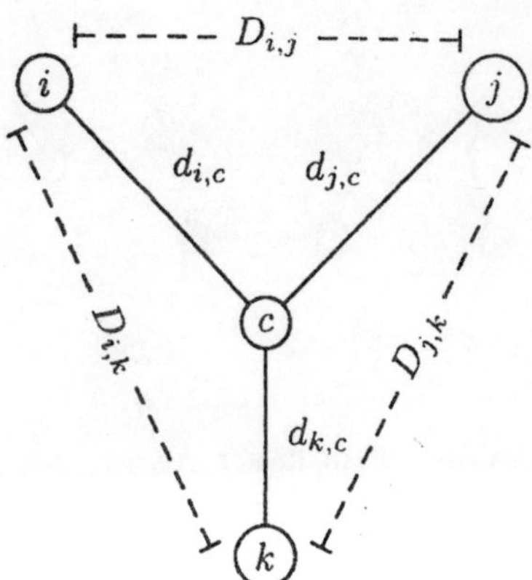

Fig. 7.8. A tree with three leaves.

leaves i and j. There are many ways to generate distance matrices: for example, one can sequence a particular gene in n species and define $D_{i,j}$ as the edit distance between this gene in species i and species j.

It is not difficult to construct a tree that fits any given 3×3 (symmetric non-negative) matrix D. This binary unrooted tree has four vertices i, j, k as leaves and vertex c as the center. The lengths of each edge in the tree are defined by the following three equations with three variables $d_{i,c}$, $d_{j,c}$, and $d_{k,c}$:

$$d_{i,c} + d_{j,c} = D_{i,j} \qquad d_{i,c} + d_{k,c} = D_{i,k} \qquad d_{j,c} + d_{k,c} = D_{j,k}.$$

The solution is given by

$$d_{i,c} = \frac{D_{i,j} + D_{i,k} - D_{j,k}}{2} \quad d_{j,c} = \frac{D_{j,i} + D_{j,k} - D_{i,k}}{2}$$

$$d_{k,c} = \frac{D_{k,i} + D_{k,j} - D_{i,j}}{2}$$

An unrooted binary tree with n leaves has $2n - 3$ edges, so fitting a *given* tree to an $n \times n$ distance matrix D leads to solving a system of $\binom{n}{2}$ equations with $2n - 3$ variables. For $n = 4$ this amounts to solving six equations with only five variables. Of course, it is not always possible to solve this system, making it hard or impossible to construct a tree from D. A matrix $(D_{i,j})$ is called *additive* if there exists a tree T with $d_{i,j}(T) = D_{i,j}$, and *nonadditive* otherwise.

The Distance-Based Phylogeny problem may not have a solution, but if it does—that is, if D is additive—there exists a simple algorithm to solve it. We emphasize the fact that we are somehow given the

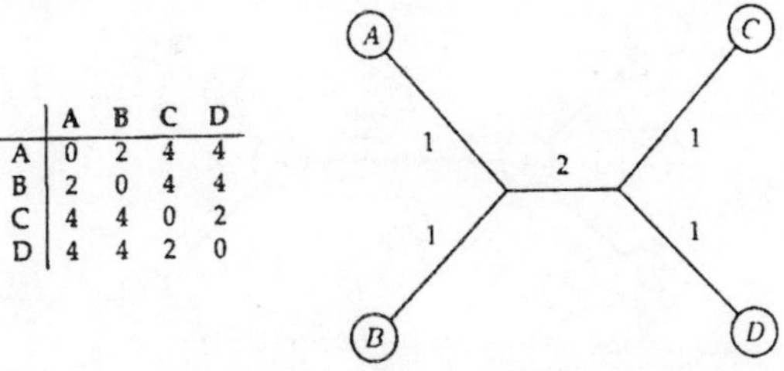

	A	B	C	D
A	0	2	4	4
B	2	0	4	4
C	4	4	0	2
D	4	4	2	0

(a) Additive matrix and the corresponding tree

	A	B	C	D
A	0	2	2	2
B	2	0	3	2
C	2	3	0	2
D	2	2	2	0

?

(b) Non-additive matrix

Fig. 7.9. Additive and nonadditive matrices.

matrix of evolutionary distances between each pair of species, and we are searching for both the shape of the tree that fits this distance matrix and the weights for each edge in the tree.

Reconstructing Trees from Additive Matrices

A "simple" way to solve the Distance-Based Phylogeny problem for additive trees is to find a pair of *neighbouring* leaves, that is, leaves that have the same parent vertex. Figure 7.10 illustrates that for a pair of neighbouring leaves i and j and their parent vertex k, the following equality holds for every other leaf m in the tree:

$$D_{k,m} = \frac{D_{i,m} + D_{j,m} - D_{i,j}}{2}$$

Therefore, as soon as a pair of neighbouring leaves i and j is found, one can remove the corresponding rows and columns i and j from the distance matrix and add a new row and column corresponding to their parent k. Since the distance matrix is additive, the distances from k to other leaves are recomputed as

$$D_{k,m} = \frac{D_{i,m} + D_{j,m} - D_{i,j}}{2}.$$

This transformation leads to a simple algorithm for the Distance-Based Phylogeny problem that finds a pair of neighbouring leaves and reduces

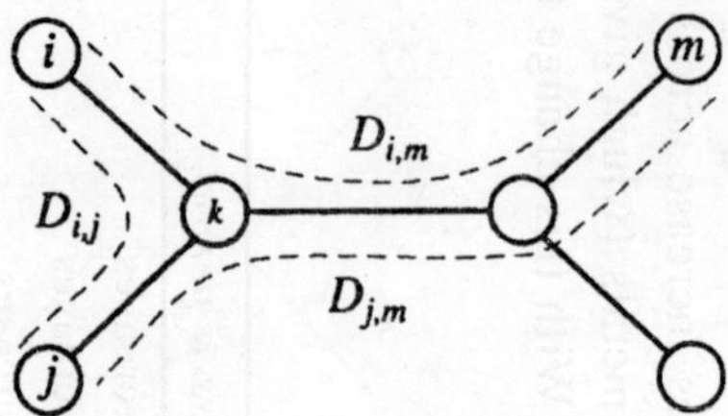

Fig. 7.10. If i and j are neighbouring leaves and k is their parent, then $D_{k,m}=[(D_{i,m}+D_{j,m}-D_{i,j})/2]$ for every other vertex m in the tree.

the size of the tree at every step. One problem with the-described approach is that it is not very easy to find neighbouring leaves! One might be tempted to think that a pair of closest leaves (i.e., the leaves i and j with minimum $D_{i,j}$) would represent a pair of neighbouring leaves, but a glance will show that this is not true. Since finding neighbouring leaves using only the distance matrix is a nontrivial problem, we postpone exploring this until the next section and turn to another approach.

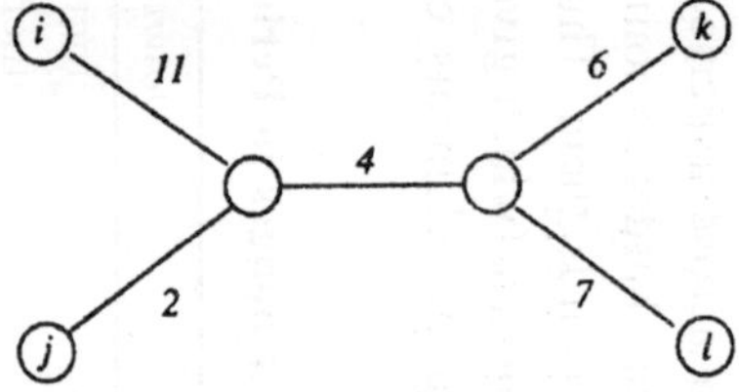

Fig. 7.11. The two closest leaves (j and k) are not neighbours in the tree.

Figure 7.12 illustrates the process of shortening *all* "hanging" edges of a tree T, that is, edges leading to leaves. If we reduce the length of every hanging edge by the same small amount δ, then the distance matrix of the resulting tree is $(d_{i,j} - 2\delta)$ since the distance between any two leaves is reduced by 2δ. Sooner or later this process will lead to "collapsing" one of the leaves when the length of the corresponding hanging edge becomes equal to 0 (when δ is equal to the length of the shortest hanging edge). At this point, the original tree $T = T_n$ with n leaves will be transformed into a tree T_{n-1} with $n - 1$ or fewer leaves.

Although the distance matrix D does not explicitly contain information about δ, it is easy to derive both δ and the location of the collapsed leaf in T_{n-1} by the method described below. Thus, one can perform a series of tree transformations $T_n \rightarrow T_{n-1} \rightarrow \dots \rightarrow T_3 \rightarrow T_2$, then construct the tree T_2 (which is easy, since it consists of only a single edge), and then perform a series of reverse transformations T_2

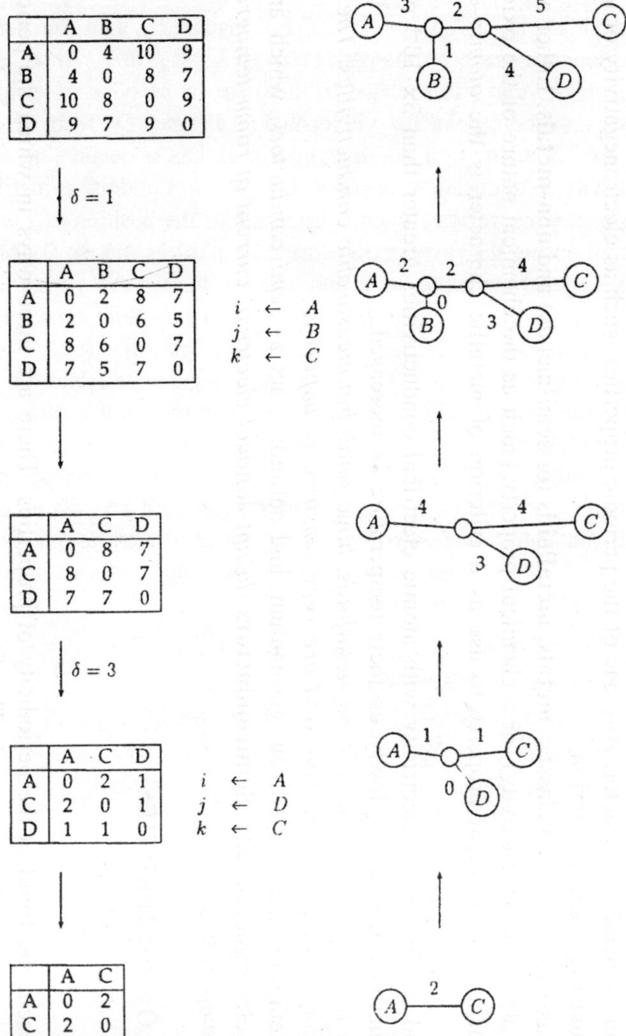

	A	B	C	D
A	0	4	10	9
B	4	0	8	7
C	10	8	0	9
D	9	7	9	0

	A	B	C	D
A	0	2	8	7
B	2	0	6	5
C	8	6	0	7
D	7	5	7	0

	A	C	D
A	0	8	7
C	8	0	7
D	7	7	0

	A	C	D
A	0	2	1
C	2	0	1
D	1	1	0

	A	C
A	0	2
C	2	0

Fig. 7.12. The iterative process of shortening the hanging edges of a tree.

$\rightarrow T_3 \rightarrow \ldots \rightarrow T_{n-1} \rightarrow T_n$ recovering information about the collapsed edges at every step.

A triple of distinct elements $1 \leq i, j, k \leq n$ is called *degenerate* if $D_{i,j} + D_{j,k} = D_{i,k}$, which is essentially just an indication that j is located on the path from i to k in the tree. If D is additive, $D_{i,j} + D_{j,k} \geq D_{i,k}$ for every triple i, j, k. We call the entire matrix D degenerate if it has at least one degenerate triple. If (i, j, k) is a degenerate triple, and some tree T fits matrix D, then the vertex j should lie somewhere on the path between i and k in T. Another way to state this is that j is attached to this path by an edge of weight 0, and the attachment point for j is located at distance $D_{i,j}$ from vertex i. Therefore, if an $n \times n$ additive matrix D has a degenerate triple, then it will be reduced to an $(n - 1) \times (n - 1)$ additive matrix by simply excluding vertex j from consideration; the position of j will be recovered during the reverse transformations. If the matrix D does not have a degenerate triple, one can start reducing the values of all elements in D by the same amount 2δ until the point at which the distance matrix becomes degenerate for the first time (i.e., δ is the minimum value for which $(D_{i,j} - 2\delta)$ has a degenerate triple for some i and j). Determining how to calculate the minimum value of δ (called the ***trimming parameter***) is left as a problem at the end of this chapter. Though you do not have the tree T, this operation corresponds to shortening all of the hanging edges in T by δ until one of the leaves ends up on the evolutionary path between two other leaves for the first time. This intuition motivates the following recursive algorithm for finding the tree that fits the data.

AdditivePhylogeny(D)

```
if D is a 2 × 2 matrix
  T ← the tree consisting of a single edge of length D_{1,2}.
  return T
if D is non-degenerate
  δ ← trimming parameter of matrix D
  for all 1 ≤ i ≠ j ≤ n
    D_{i,j} ← D_{i,j} - 2δ
else
  δ ← 0
Find a triple i, j, k in D such that D_{ij} + D_{jk} = D_{ik}
x ← D_{i,j}
Remove jth row and jth column from D.
T ← AdditivePhylogeny(D)
Add a new vertex υ to T at distance x from i to k
Add j back to T by creating an edge (υ, j) of length 0
for every leaf l in T
  if distance from l to υ in the tree T does not equal D_{l,j}
    output "Matrix D is not additive"
```

return

Extend hanging edges leading to all leaves by δ

return T

The *AdditivePhylogeny* algorithm above provides a way to check if the matrix D is additive. While this algorithm is intuitive and simple, it is not the most efficient way to construct additive trees. Another way to check additivity is by using the following "four-point condition". Let $1 \le i,j,k,l \le n$ be four distinct indices. Compute 3 sums: $D_{i,j} + D_{k,l}$, $D_{i,k} + D_{j,l}$, and $D_{i,l} + D_{j,k}$. If D is an additive matrix then these three sums can be represented by a tree with four leaves. Moreover, two of these sums represent the same number (the sum of lengths of all edges in the tree plus the length of the middle edge) while the third sum represents another smaller number (the sum of lengths of all edges in the tree minus the length of the middle edge). We say that elements $1 \le i,j,k,l \le n$ satisfy the *four-point condition* if two of the sums $D_{i,j} + D_{k,l}$, $D_{i,k} + D_{j,l}$, and $D_{i,l} + D_{j,k}$ are the same, and the third one is smaller than these two.

If the distance matrix D is not additive, one might want instead to find a tree that approximates D using the sum of squared errors $\Sigma_{i,j}(d_{i,j}(T) - D_{i,j})^2$ as a measure of the quality of the approximation. This leads to the (*NP*-hard) Least Squares Distance-Based Phylogeny problem.

Evolutionary Trees and Hierarchical Clustering

Biologists often use variants of hierarchical clustering to construct evolutionary trees. UPGMA (*Unweighted Pair Group Method with Arithmetic Mean*) is a particularly simple clustering algorithm. The UPGMA algorithm is a variant of *HierarchicalClustring* that uses a different approach to compute the distance between clusters, and assigns *heights* to vertices of the constructed tree. Thus, the length of an edge (u, υ) is defined to be the difference in heights of the vertices υ and u. The height plays the role of the molecular clock, and allows one to "date" the divergence point for every vertex in the evolutionary tree.

Given clusters C_1 and C_2, UPGMA defines the distance between them to be the average pairwise distance:

$$D(C_1, C_2) = \frac{1}{|C_1||C_2|} \sum_{i \in C_1} \sum_{j \in C_2} D(i,j).$$

At heart, UPGMA is simply another hierarchical clustering algorithm that "dates" vertices of the constructed tree.

UPGMA (D, n)

Form n clusters, each with a single element
Construct a graph T by assigning an isolated vertex to each cluster
Assign height $h(\upsilon) = 0$ to every vertex υ in this graph
while there is more than one cluster
    Find the two closest clusters C_1 and C_2
    Merge C_1 and C_2 into a new cluster C with $|C_1| + |C_2|$ elements
    for every cluster $C^* \neq C$
        $D(C, C^*) = \frac{1}{|C| \cdot |C^*|} \sum_{i \in C} \sum_{j \in C^*} D(i, j)$
    Add a new vertex C to T and connect to vertices C_1 and C_2.
    $h(C) \leftarrow \frac{D(C_1, C_2)}{2}$
    Assign length $h(C) - h(C_1)$ to the edge (C_1, C)
    Assign length $h(C) - h(C_2)$ to the edge (C_2, C)
    Remove rows and columns of D corresponding to C_1 and C_2
    Add a row and column to D for the new cluster C
return T

UPGMA produces a special type of rooted tree that is known as *ultrametric*. In ultrametric trees the distance from the root to any leaf is the same.

We can now return to the "neighbouring leaves" idea that we developed and then abandoned in the previous section. In. 1987 Naruya Saitou and Masatoshi Nei developed an ingenious *neighbour joining* algorithm for phylogenetic tree reconstruction. In the case of additive trees, the neighbour joining algorithm somehow magically finds pairs of neighbouring leaves and proceeds by substituting such pairs with the leaves' parent. However, neighbour joining works well not only for additive distance matrices but for many others as well: it does not assume the existence of a molecular clock and ensures that the clusters that are merged in the course of tree reconstruction are not only close to each other (as in UPGMA) but also are far apart from the rest.

For a cluster C, define

$$u(C) = \frac{1}{\text{number of cluster} - 2} \sum_{\text{all clusters } C'} D(C, C')$$

as a measure of the separation of C from other clusters. To choose which two clusters to merge, we look for the clusters C_1 and C_2 that are simultaneously close to each other and far from others. One may try to merge clusters that simultaneously minimize $D(C_1, C_2)$ and maximize $u(C_1) + u(C_2)$. However, it is unlikely that a pair of clusters

C_1 and C_2 that simultaneously minimize $D(C_1, C_2)$ and maximize $u(C_1) + u(C_2)$ exists. As an alternative, one opts to minimize $D(C_1, C_2) - u(C_1) - u(C_2)$. This approach is used in the *NeighbourJoining* algorithm below.

NeighbourJoining (D, n)

```
Form n clusters, each with a single element
Construct a graph T by assigning an isolated vertex to each cluster
while there is more than one cluster
    Find clusters C1 and C2 minimizing D(C1, C2) - u(C1) - u(C2)
    Merge C1 and C2 into a new cluster C with |C1|+|C2| elements
    Compute D(C,C*) = (D(C1,C) + D(C2,C)) / 2 to every other cluster C*
    Add a new vertex C to T and connect it to vertices C1 and C2
    Assign length ½D(C1,C2) + ½(u(C1) - u(C2)) to the edge (C1, C)
    Assign length ½D(C1,C2) + ½(u(C2) - u(C1)) to the edge (C2, C)
    Remove rows and columns of D corresponding to C1 and C2
    Add a row and column to D for the new cluster C
return T
```

Character-Based Tree Reconstruction

Evolutionary tree reconstruction often starts by sequencing a particular gene in each of n species. After aligning these genes, biologists end up with an $n \times m$ alignment matrix (n species, m nucleotides in each) that can be further transformed into an $n \times n$ distance matrix. Although the distance matrix could be analyzed by distance-based tree reconstruction algorithms, a certain amount of information gets lost in the transformation of the alignment matrix into the distance matrix, rendering the reverse transformation of distance matrix back into the alignment matrix impossible. A better technique is to use the alignment matrix directly for evolutionary tree reconstruction. *Character-based* tree reconstruction algorithms assume that the input data are described by an $n \times m$ matrix (perhaps an alignment matrix), where n is the number of species and m is the number of *characters*. Every row in the matrix describes an existing species and the goal is to construct a tree whose leaves correspond to the n existing species and whose internal vertices correspond to ancestral species. Each internal vertex in the tree is labeled with a character string that describes, for example, the hypothetical number of legs in that ancestral species. We want to determine what character strings at internal nodes would best explain the character strings for the *n* observed species.

The use of the word "character" to describe an attribute of a species is potentially confusing, since we often use the word to refer to letters from an alphabet. We are not at liberty to change the terminology that biologists have been using for at least a century, so for the next section we will refer to nucleotides as *states* of a character. Another possible character might be "number of legs," which is not very informative for mammalian evolutionary studies, but could be somewhat informative for insect evolutionary studies.

An intuitive score for a character-based evolutionary tree is the total number of mutations required to explain all of the observed

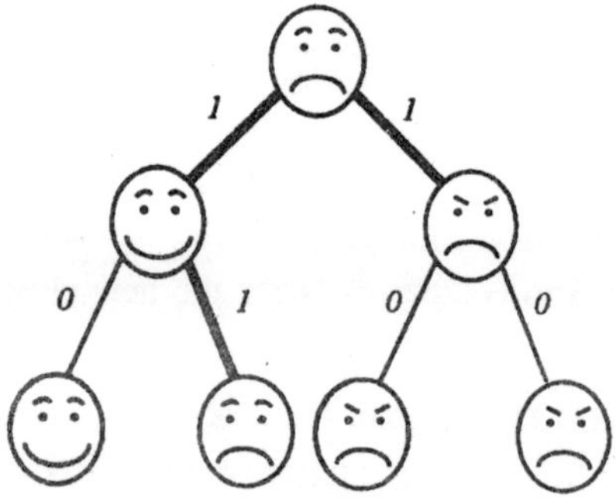

(a) *Parsimony Score=3*

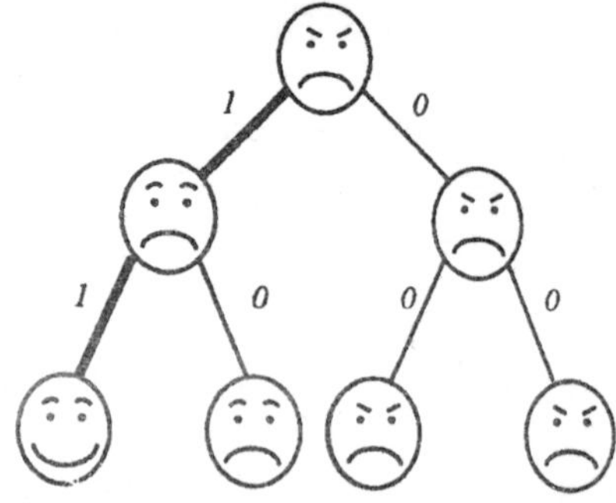

(b) *Parsimony Score=2*

Fig. 7.13. If we label a tree's leaves with characters, and choose labels for each internal vertex, we implicitly create a parsimony score for the tree. By changing the labels in (a) we are able to create a tree with a better parsimony score in (b).

character sequences. The *parsimony* approach attempts to minimize this score, and follows the philosophy of Occam's razor: find the simplest explanation of the data.

Given a tree T with every vertex labeled by an m-long string of characters, one can set the length of an edge (υ, w) to the Hamming distance $d_H(\upsilon, w)$ between the character strings for υ and w. The *parsimony score* of a tree T is simply the sum of lengths of its edges $\sum_{\text{All edges } (\upsilon,w) \text{ of the tree}} d_H(\upsilon, w)$. In reality, the strings of characters assigned to internal vertices are unknown and the problem is to find strings that minimize the parsimony score.

Two particular incarnations of character-based tree reconstruction are the Small Parsimony problem and the Large Parsimony problem. The Small Parsimony problem assumes that the tree is given but the labels of its internal vertices are unknown, while the vastly more difficult Large Parsimony problem assumes that neither the tree structure nor the labels of its internal vertices are known.

Small Parsimony Problem

An attentive reader should immediately notice that, because the characters in the string are independent, the Small Parsimony problem can be solved independently for each character. Therefore, to devise an algorithm, we can assume that every leaf is labeled by a single character rather than by a string of m characters.

As we have seen in previous chapters, sometimes solving a more general—and seemingly more difficult—problem may reveal the solution to the more specific one. In the case of the Small Parsimony Problem we will first solve the more general *Weighted Small Parsimony* problem, which generalizes the notion of parsimony by introducing a scoring matrix. The length of an edge connecting vertices υ and w in the Small Parsimony problem is defined as the Hamming distance, $d_H(\upsilon,w)$, between the character strings for υ and w. In the case when every leaf is labeled by a single character in a k-letter alphabet, $d_H(\upsilon,w) = 0$ if the characters corresponding to υ and w are the same, and $d_H(\upsilon,w) = 1$ otherwise. One can view such a scoring scheme as a $k \times k$ scoring matrix $(\delta_{i,j})$ with diagonal elements equal to 0 and all other elements equal to 1. The Weighted Small Parsimony problem simply assumes that the scoring matrix $(\delta_{i,j})$ is an arbitrary $k \times k$ matrix and minimizes the weighted parsimony score $\sum_{\text{all edge } (\upsilon,w) \text{ in the tree}} \delta_{\upsilon,w}$.

In 1975 David Sankoff came up with the following dynamic programming algorithm for the Weighted Small Parsimony problem. As usual in dynamic programming, the Weighted Small Parsimony

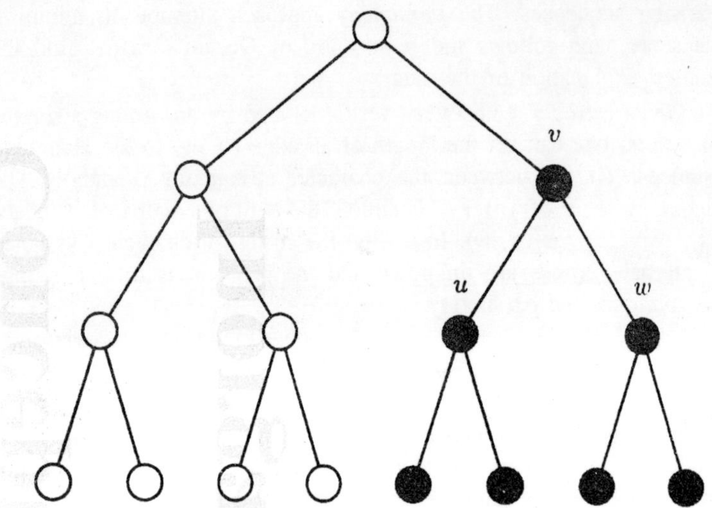

Fig. 7.14. A subtree of a larger tree. The shaded vertices form a tree rooted at the topmost shaded node.

problem for T is reduced to solving the Weighted Small Parsimony Problem for smaller subtrees of T. As we mentioned earlier, a rooted tree can be viewed as a directed tree with all of its edges directed away from the root toward the leaves. Every vertex υ in the tree T defines a *subtree* formed by the vertices beneath υ, which are all of the vertices that can be reached from υ. Let $s_t(\upsilon)$ be the minimum parsimony score of the subtree of υ under the assumption that vertex υ has character t. For an internal vertex υ with children u and w, the score $s_t(\upsilon)$ can be computed by analyzing k scores $s_i(u)$ and k scores $s_i(w)$ for $1 \le i \le k$ (below, i and j are characters):

$$s_t(\upsilon) = \min_i\{s_i(u) + \delta_{i,t}\} + \min_j\{s_j(w) + \delta_{j,t}\}$$

The initial conditions simply amount to an assignment of the scores $s_t(\upsilon)$ at the leaves according to the rule: $s_t(\upsilon) = 0$ if υ is labeled by letter t and $s_t(\upsilon) = \infty$ otherwise. The minimum weighted parsimony score is given by the smallest score at the root, $s_t(r)$. Given the computed values $s_t(\upsilon)$ at all of the vertices in the tree, one can reconstruct an optimal assignment of labels using a backtracking approach. The running time of the algorithm is $O(nk)$.

In 1971, even before David Sankoff solved the Weighted Small Parsimony problem, Walter Fitch derived a solution of the (unweighted) Small Parsimony problem. The Fitch algorithm below is essentially dynamic programming in disguise. The algorithm assigns a set of letters

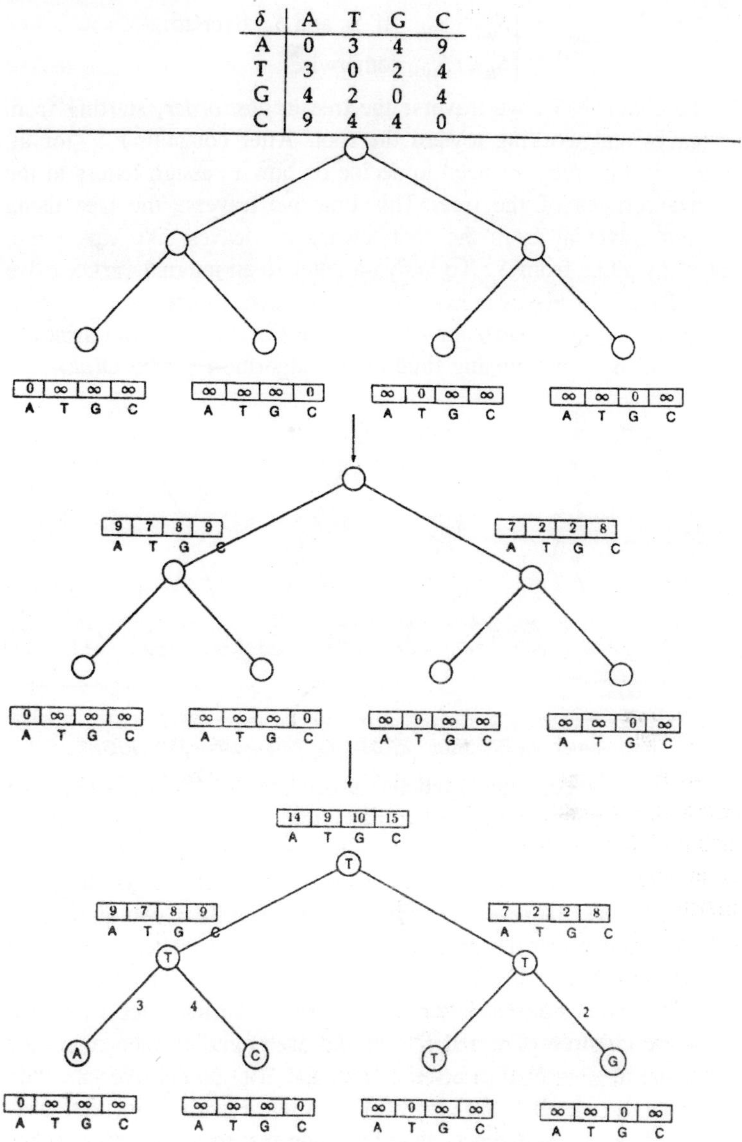

δ	A	T	G	C
A	0	3	4	9
T	3	0	2	4
G	4	2	0	4
C	9	4	4	0

Fig. 7.15. An illustration of Sankoff's algorithm.

S_υ to every vertex in the tree in the following manner. For each leaf υ, S_υ consists of single letter that is the label of this leaf. For any internal vertex υ with children u and w, S_υ is computed from the sets S_u and S_w according to the following rule:

$$S_\upsilon = \begin{cases} S_u \cap S_w, & \text{if } S_u \text{ and } S_w \text{ overlap} \\ S_u \cup S_w, & \text{otherwise} \end{cases}$$

To compute S_υ, we traverse the tree in *post-order*, starting from the leaves and working toward the root. After computing S_υ for all vertices in the tree, we need to decide on how to assign letters to the internal vertices of the tree. This time we traverse the tree using *preorder* traversal from the root toward the leaves. We can assign root *r* any letter from S_r. To assign a letter to an internal vertex υ we check if the (already assigned) label of its parent belongs to S_υ. If yes, we choose the same label for υ; otherwise we label υ by an arbitrary letter from S_υ. The running time of this algorithm is also O(*nk*).

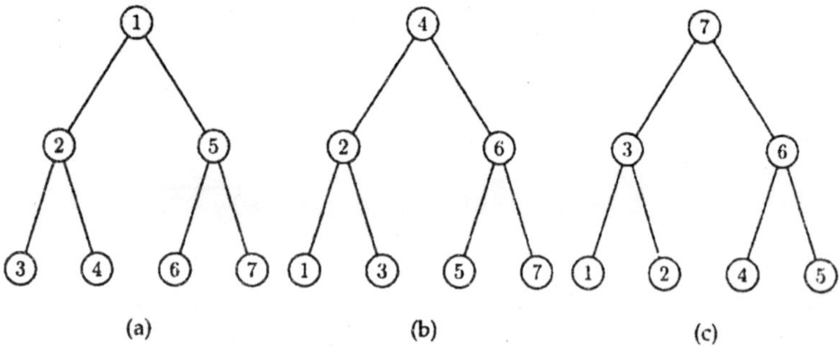

Fig. 7.16. Three methods of traversing a tree. (a) Pre-order: SELF, LEFT, RIGHT. (b) In-order: LEFT, SELF, RIGHT. (c) Post-order: LEFT, RIGHT, SELF.

At first glance, Fitch's labeling procedure and Sankoff's dynamic programming algorithm appear to have little in common. Even though Fitch probably did not know about application of dynamic programming for evolutionary tree reconstruction in 1971, the two algorithms are almost identical. To reveal the similarity between these two algorithms let us return to Sankoff's recurrence. We say that character *t* is optimal for vertex υ if it yields the smallest score, that is, if $s_t(\upsilon) = \min_{1 \le i \le k} s_i(\upsilon)$. The set of optimal letters for a vertex υ forms a set $S(\upsilon)$. If *u* and *w* are children of υ and if $S(u)$ and $S(w)$ overlap, then it is easy to see that $S(\upsilon) = S(u) \cap S(w)$. If $S(u)$ and $S(w)$ do not overlap, then it is easy to see that $S(\upsilon) = S(u) \cup S(w)$. Fitch's algorithm uses exactly the same recurrence, thus revealing that these two approaches are algorithmic twins.

Large Parsimony Problem

Not surprisingly, the Large Parsimony problem is *NP*-complete. In the case *n* is small, one can explore all tree topologies with *n*

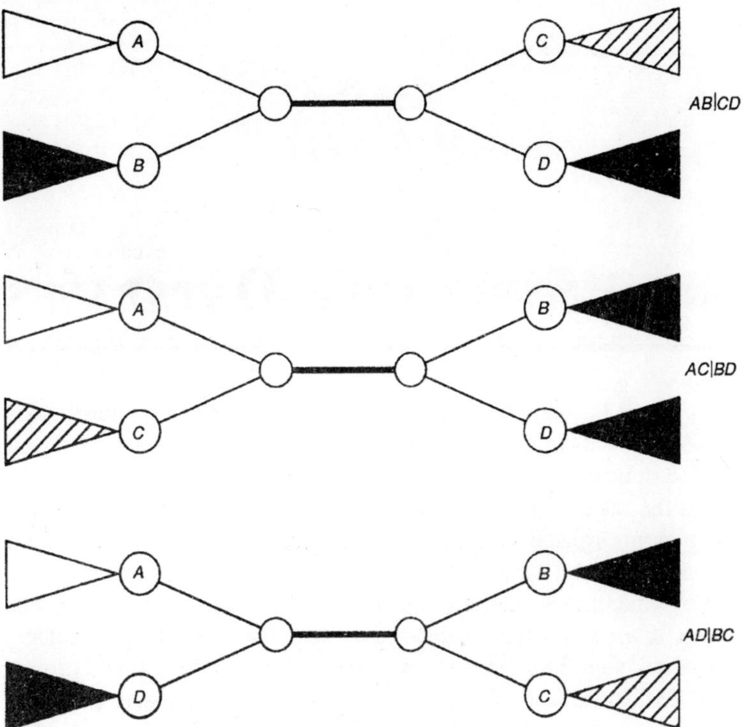

Fig. 7.17. Three ways of combining the four subtrees defined by an internal edge.

leaves, solve the Small Parsimony problem for each topology, and select the best one. However, the number of topologies grows very fast with respect to n, so biologists often use local search heuristics (e.g., greedy algorithms) to navigate in the space of all topologies. *Nearest neighbour interchange* is a local search heuristic that defines neighbours in the space of all trees. Every *internal* edge in a tree defines four subtrees A, B, C, and D that can be combined into a tree in three different ways that we denote $AB|CD$, $AC|BD$, and $AD|BC$. These three trees are called ***neighbours*** under the nearest neighbour interchange transformation. A greedy approach to the Large Parsimony problem is to start from an arbitrary tree and to move (by nearest neighbour interchange) from one tree to another if such a move provides the best improvement in the parsimony score among all neighbours of the tree T.

8

STRUCTURE DATABASES

This chapter introduces biomolecular structures from a bioinformatics perspective, with special emphasis on the sequences that are contained in three-dimensional structures. The major goal of this chapter is to inform the reader about the contents of structure database records and how they are treated, and sometimes mistreated, by popular software programs. This chapter does not cover the computational processes used by structural scientists to obtain three-dimensional structures, nor does it discuss the finer points of comparative protein architecture. Several excellent monographs regarding protein architecture and protein structure determination methods are already widely available and often found, in campus bookstores.

The imagery of protein and nucleic acid structures has become a common feature of biochemistry textbooks and research articles. This imagery can be beautiful and intriguing enough to blind us to the experimental details an image represents—the underlying biophysical methods and the effort of hard-working X-ray crystallographers and *nuclear magnetic resonance* (NMR) spectroscopists. The data stored in structure database records represents a practical summary of the experimental data. It is, however, not the data gathered directly by instruments, nor is it a simple mathematical transformation of that data. Each structure database record carries assumptions and biases that change as the stale of the art in structure determination advances. Nevertheless, each biomolecular structure is a hard-won piece of crucial information and provides potentially critical information regarding the function of any given protein sequence.

Since the first edition of this book was released, the software for viewing and manipulating three-dimensional structures has improved dramatically. Another major change has come in the form of an organizational transition, with the Protein Data Bank (PDB) moving from the Brookhaven National Laboratories to the Research Collaboratory for Structural Biology. The result has been a complete change in the organization of the PDB web site. The impact of these changes for biologists will be discussed herein.

Notion of Three-Dimensional Molecular Structure Data

Let us begin with a mental exercise in recording the three-dimensional data of a biopolymer. Consider how we might record, on paper, all the details and dimensions of a three-dimensional ball-and-stick model of a protein like myoglobin. One way to begin is with the sequence, which can be obtained by tracing out the backbone of the three-dimensional model. Beginning from the NH_2-terminus, we identify each amino acid side chain by comparing the atomic structure of each residue with the chemical structure of the 20 common amino acids, possibly guided by an illustration of amino acid structures from a textbook. Once the sequence has been written down, we proceed with making a two-dimensional sketch of the biopolymer with all its atoms, element symbols, and bonds, possibly taking up several pieces of paper. The same must be done for the heme ligand, which is an important functional part of the myoglobin molecule. After drawing its chemical structure on paper, we might record the three-dimensional data by measuring the distance of each atom in the model starting from some origin point, along some orthogonal axis system. This would provide the *x*-, *y*-, and *z*-axis distances to each atomic "ball" in the ball-and-stick structure.

The next step is to come up with a bookkeeping scheme to keep all the (x, y, z) coordinate information connected to the identity of each atom. The easiest approach may be to write the (x, y, z) value as a coordinate triple on the same pieces of paper used for the two-dimensional sketch of the biopolymer, right next to each atom. This associates the (x, y, z) value with the atom it is attached to.

This mental exercise helps to conceptualize what a three-dimensional structure database record ought to contain. There are two things that have been recorded here: the chemical structure and the locations of the individual atoms in space. This is an adequate "human-readable" record of the structure, but one probably would not expect a computer to digest it easily. The computer needs clear encoding of the associations of atoms, bonds, coordinates, residues, and molecules, so that one may construct software that can read the data in an unambiguous manner. Here is where the real exercise in structural bioinformatics begins.

Coordinates, Sequences, and Chemical Graphs

The most obvious data in a typical three-dimensional structure record, regardless of the file format in use, is the *coordinate data*, the locations in space of the atoms of a molecule. These data are represented by (x, y, z) triples, distances along each axis to some arbitrary origin in space. The coordinate data for each atom is attached to a list of labeling information in the structure record: which element, residue, and molecule each point in space belongs to. For the standard biopolymers (DNA, RNA, and proteins), this labeling information can be derived starting with the raw sequence. Implicit in each sequence is considerable chemical data. We can infer the complete chemical connectivity of the biopolymer molecule directly from a sequence, including all its atoms and bonds, and we could make a sketch, just like the one described earlier, from sequence information alone. We refer to this "sketch" of the molecule as the *chemical graph* component of a three-dimensional structure. Every time a sequence is presented in this book or elsewhere, remember that it can encode a fairly complete description of the chemistry of that molecule.

When we sketch all the underlying atoms and bonds representing a sequence, we may defer to a textbook showing the chemical structures of each residue, lest we forget a methyl group or two. Likewise, computers could build up a sketch like a representation of the chemical graph of a structure in memory using a *residue dictionary*, which contains a table of the atom types and bond information for each of the common amino acid and nucleic acid building blocks. What sequence is unable to encode is information about posttranslational modifications. For example, in the structure databases, a phosphorylated tyrosine residue is indicated as "X" in the one letter code—essentially an unknown! Any residue that has had an alteration to its standard chemical graph will, unfortunately, be indicated as X in the one-letter encoding of sequence.

Atoms, Bonds, and Completeness

Molecular graphics visualization software performs an elaborate "connect-the-dots" process to make the wonderful pictures of protein structure, like the structure for insulin. The connections used are, of course, the chemical bonds between all the atoms. In current use, three-dimensional molecular structure database records employ two different "minimalist" approaches regarding the storage of bond data.

The original approach to recording atoms and bonds is something we shall call the *chemistry rules* approach. The rules are the observable physical rules of chemistry, such as, "the average length of a stable C—C bond is about 1.5 angstroms." Applying these rules to derive the bonds means that any two coordinate locations in space that are 1.5 Å apart and are tagged as carbon atoms always form a single bond. With the chemistry rules approach, we can simply disregard the bonds. A perfect and complete structure can be recorded without any bond information, provided it does not break any of the rules of chemistry in atomic locations. Obviously, this is not always the case, and specific examples of this will be presented later in this chapter.

The chemistry rules approach ended up being the basis for the original three-dimensional biomolecular structure file format, the PDB format from the Protein Data Bank at Brookhaven. These records, in general, lack complete bond information for biopolymers. The working assumption is that no residue dictionary is required for interpretation of data encoded by this approach, just a table of bond lengths and bond types for every conceivable pair of bonded atoms is required.

Every software package that reads in PDB data files must reconstruct the bonds based on these rules. However, the rules we are describing have never been explicitly codified for programmers. This means that interpreting the bonding in PDB files is left for the *programmer* to decide, and, as a result, software can be inconsistent in the way it draws bonds, especially when different algorithms and distance tolerances are used. The PDB file approach is minimalist in terms of the data stored in a record, and deciphering it often requires much more sophisticated logic than would be needed if the bonding information and chemical graph were explicitly specified in the record. Rarely is this logic properly implemented, and it may in fact be impossible to deal with all the exceptions in the PDB file format. Each exception to the bonding rules needs to be captured by complicated logic statements programmed on a case-by-case basis.

The second approach to describing a molecule is what we call the *explicit bonding approach*, the method that is used in the database records of the Molecular Modeling Database (MMDB), which is, in turn, derived from the data in PDB. In the MMDB system, the data file contains all of its own explicit bonding information. MMDB uses a standard residue dictionary, a record of all the atoms and bonds in the polymer forms of amino acid and nucleic acid residues, plus end-terminal variants. Such data dictionaries are common in the specialized software used by scientists to solve X-ray or NMR structures. The software that reads in MMDB data can use the bonding information supplied in the dictionary to connect atoms together, without trying to enforce (or force) the rules of chemistry. As a result, the three-dimensional coordinate data are consistently interpreted by visualization software, regardless of type. This approach also lends itself to inherently simpler software, because exceptions to bonding rules are recorded within the database file itself and read in without the need for another layer of exception-handling codes.

Scientists that are unfamiliar with structure data often expect all structures in the public databases to be of "textbook" quality. They are often surprised when parts of a structure are missing. The availability of a three-dimensional database record for a particular molecule does not ever imply its completeness. Structural completeness is strictly defined as follows: *At least one coordinate value for each and every atom in the chemical graph is present.*

Structural completeness is quite rare in structure database records. Most X-ray structures lack coordinates for hydrogen atoms because the locations of hydrogens in space are not resolved by the experimental methods currently available. However, some modeling software can be used to predict the locations of these hydrogen atoms and reconstruct a structure record with the now-modeled hydrogens added. It is easy to identify the products of molecular modeling in structure databases. These often have overly complete coordinate data, usually with all possible hydrogen atoms present that could not have been found using an experimental method.

PBD: Protein Data Bank at the Research Collaboratory For Structural Bioinformatics (RCSB)

The use of computers in biology has its origins in biophysical methods, such as X-ray crystallography. Thus, it is not surprising that the first "bioinformatics" database was built to store complex three-dimensional data. The Protein Data Bank, originally developed and housed at the Brookhaven National Laboratories, is now managed and maintained by the Research Collaboratory for Structural Bioinformatics (RCSB). RCSB is a collaborative effort involving scientists at the San Diego Supercomputing Center, Rutgers University, and the National Institute of Standards and Technology. The collection contains all publicly available three-dimensional structures of proteins, nucleic acids, carbohydrates, and a variety of other complexes experimentally determined by X-ray crystallographers and NMR spectroscopists. This section focuses briefly on the database and bioinformatics services offered through RCSB.

RCSB Database Services

The World Wide Web site of the Protein Data Bank at the RCSB offers a number of services for submitting and retrieving three-dimensional structure data. The home page of the RCSB site provides links to services for depositing three-dimensional structures, information on how to obtain the status of structures undergoing processing for submission, ways to download the PDB database, and links to other relevant sites and software.

PDB Query and Reporting

Starting at the RCSB home page, one can retrieve three-dimensional structures using two different query engines. The SearchLite system is the one most often used, providing text searching across the database. The SearchFields interface provides the additional ability to search specific fields within the database. Both of these systems report structure matches to the query in the form of Structure Summary pages. The RCSB Structure Summary page links are to other Web pages that themselves provide a large number of links, and it may be confusing to a newcomer to not only sift through all this information but to decide which information sources are the most relevant ones for biological discovery.

Submitting structures

For those who wish to submit three-dimensional structure information to PDB, the RCSB offers its ADIT service over the Web. This service provides a data format check and can create automatic validation reports that provide diagnostics as to the quality of the structure, including bond distances and angles, torsion angles, nucleic acid comparisons, and crystal packing. Nucleic acid structures are accepted for deposition at NDB, the Nucleic Acids Database.

It has been the apparent working policy of PDB to reject three-dimensional structures that result from computational three-dimensional modeling procedures rather than from an actual physical experiment; submitting data to the PDB from a nonexperimental computational modeling exercise is strongly discouraged.

PDB-ID codes

The structure record accessioning scheme of the Protein Data Bank is a unique four-character alphanumeric code, called a PDB-ID or PDB code. This scheme uses the digits 0 to 9 and the uppercase letters A to Z. This allows for over 1.3 million possible combinations and entries. Many older records have mnemonic names that make the structures easier to remember, such as 3INS, the record for insulin shown earlier. A different method is now being used to assign PDB-IDs, with the use of mnemonics apparently being abandoned.

Database searching, PDB file retrieval, mmCIF file retrieval, and links

PDB's search engine, the Structure Explorer, can be used to retrieve PDB records. The Structure Explorer is also the primary database of links to third-party annotation of PDB structure data. There are a number of links maintained in the Structure Explorer to Internet-based three-dimensional structure services on other Web sites. The Structure Explorer also provides links to special project databases maintained by researchers interested in related topics, such as structural evolution, structure-structure similarity, and protein motions. Links to visualization tool-ready versions of the structure are provided, as well as authored two-dimensional images that can be very helpful to see how to orient a three-dimensional structure for best viewing of certain features such as binding sites.

Sequences from Structure Records

PDB file-encoded sequences are notoriously troublesome for programmers to work with. Because completeness of a structure is not always guaranteed, PDB records contain two copies of the sequence information: an *explicit sequence* and an *implicit sequence*. Both are required to reconstruct the chemical graph of a biopolymer.

Explicit sequences in a PDB file are provided in lines starting with the keyword SEQRES. Unlike other sequence databases, PDB records use the three-letter amino acid code, and nonstandard amino acids are found in many PDB record sequence entries with arbitrarily chosen three-letter names. Unfortunately, PDB records seem to lack sensible, consistent rules. In the past, some double-helical nucleic acid sequence entries in PDB were specified in a 3′-to-5′ order in an entry above the complementary strand, given in 5′-to-3′ order. Although the sequences may be obvious to a user as a representation of a double helix, the 3′-to-5′ explicit sequences are nonsense to a computer. Fortunately, the NDB project has fixed many of these types of problems, but the PDB data format is still open to ambiguity disasters from the standpoint of computer readability. As an aside, the most troubling glitch is the inability to encode element type separately from the atom name. Examples of where this becomes problematic include cases where atoms in structures having FAD or NAD cofactors are notorious for being interpreted as the wrong elements, such as neptunium (NP to Np), actinium (AC to Ac), and other nonsense elements.

Because three-dimensional structures can have multiple biopolymer chains, to specify a discrete sequence, the user must provide the *PDB chain identifier*. SEQRES entries in PDB files have a chain identifier, a single uppercase letter or blank space, identifying each individual biopolymer chain in an entry. For the structure 3INS, there are two insulin molecules in the record. The 3INS record contains sequences labeled A, B, C, and D. Knowledge of the biochemistry of insulin is required to understand that protein chains A and B are in fact derived from the same gene and that a posttranslational modification cuts the proinsulin sequence into the A and B chains observed in the PDB record. This information is not recorded in a three-dimensional structure record, nor in the sequence record for that matter. A place for such critical biological information is now being made within the BIND database. The one-letter chain-naming scheme has difficulties with the enumeration of large oligomeric three-dimensional structures, such as viral capsids, as one quickly runs out of single-letter chain identifiers.

The *implicit* sequences in PDB records are contained in the embedded stereochemistry of the (*x*, *y*, *z*) data and names of each ATOM record in the PDB file. The implicit sequences are useful in resolving explicit sequence ambiguities such as the backward encoding of nucleic acid sequences or in verifying nonstandard amino acids. In practice, many PDB file viewers (such as RasMol) reconstruct the chemical graph of a protein in a PDB record using only the *implicit* sequence, ignoring the *explicit* SEQRES information. If this software then is asked to print the sequence of certain incomplete molecules, it will produce a nonphysiological and biologically irrelevant sequence. The implicit sequence, therefore, is not sufficient to reconstruct the complete chemical graph.

Consider an example in which the sequence ELVISISALIVES is represented in the SEQRES entry of a hypothetical PDB file, but the coordinate information is missing all (*x*, *y*, *z*) locations for the subsequence ISA. Software that reads the implicit sequence will often report the PDB sequence incorrectly from the chemical graph as ELVISLIVES. A test structure to determine whether software looks only at the implicit sequence is 3TS1 as shown in the Java three-dimensional structure viewer WebMol. Here, both the implicit and explicit sequences in the PDB file to the last residue with coordinates are correctly displayed.

Validating PDB Sequences

To properly validate a sequence from a PDB record, one must first derive the *implicit* sequence in the ATOM records. This is a nontrivial processing step. If the structure has gaps because of lack of completeness, there may only be a set of *implicit sequence fragments* for a given chain. Each of these fragments must be aligned to the *explicit* sequence of the same chain provided within the SEQRES entry. This treatment will produce the complete chemical graph, including the parts of the biological sequence that may be missing coordinate data. This kind of validation is done on creation of records for the MMDB and mmCIF databases.

The best source of validated protein and nucleic acid sequences in single-letter code derived from PDB structure records is NCBI's MMDB service, which is part of the Entrez system. The sequence records from our insulin example have database accessions constructed systematically and can be retrieved from the protein sequence division of Entrez using the accessions pdb|3INS|A, pdb|3INS|B, pdb|3INS|C, and pdb|3INS|D. PDB files also have references in db-xref records to sequences in the SWISS-PROT protein database. Note that the SWISSPROT sequences will not necessarily correspond to the structure, since the validation process described here is not carried out when these links are made! Also, note that many PDB files currently have ambiguously indicated taxonomy, reflecting the presence in some of three-dimensional structures of complexes of molecules that come from different Species. The PDBeast project at NCBI has incorporated the correct taxonomic information for each biopolymer found within a given structure.

MMDB: Molecular Modeling Database at NCBI

NCBI's Molecular Modeling Database (MMDB) is an integral part of NCBI's Entrez information retrieval system. It is a compilation of all the Brookhaven Protein Data Bank three-dimensional structures of biomolecules from crystallographic and NMR studies. MMDB record; are in ASN.1 format rather than in PDB-format. Despite this, PDB formatted files can also be obtained from MMDB.

By representing the data in ASN.1 format, MMDB records have value-added information compared with the original PDB entries. Additional information includes explicit chemical graph information resulting from an extensive suite of validation procedures, the addition of uniformly derived secondary structure definitions, structure domain information, citation matching to MEDLINE, and the molecule-based assignment of taxonomy to each biologically derived protein or nucleic acid chain.

Free Text Query of Structure Records

The MMDB database can be searched from the NCBI home page using Entrez. (MMDB is also referred to as the NCBI Structure division.) Search fields in MMDB include PDB and MMDB ID codes, free text from the original PDB REMARK records, author name, and other bibliographic fields. For more specific, fielded queries, the RCSB site is recommended.

MMDB Structure Summary

MMDB's Web interface provides a Structure Summary page for each MMDB structure record. MMDB Structure Summary pages provide the FASTA-formatted sequences for each chain in the structure, links to MEDLINE references, links lo the 3DBAtlas record and the Brookhaven PDB site, links to protein or nucleic acid sequence neighbours for each chain in the structure, and links to VAST structure-structure comparisons for each domain on each chain in the structure.

BLAST Against PDB Sequences: New Sequence Similarities

When a researcher wishes to find a structure related to a new sequence, NCBI's BLAST can be used because the BLAST databases contain a copy of all the validated sequences from MMDB. The BLAST Web interface can be used to perform the query by pasting a sequence in FASTA format into the sequence entry box and then selecting the "pdb" sequence database. This will yield a search against all the validated sequences in the current public structure database.

Entrez Neighbouring: Known Sequence Similarities

If one is starting with a sequence that is already in Entrez, BLAST has, in essence, already been performed. Structures that are similar in sequence to a given protein sequence can be found by means of Entrez's neighbouring facilities.

Structure File Formats

PDB

The PDB file format is column oriented, like that of the punched cards used by early FORTRAN programmers. The exact file format specification is available through the PDB Web site. Most software developed by structural scientists is written in FORTRAN, whereas the rest of the bioinformatics world has adopted other languages, such as those based on C. PDB files are often a paradox: they look rather easy to parse, but they have a few nasty surprises, as already alluded to in this chapter. To the uninitiated, the most obvious problem is that the information about biopolymer bonds is missing, obliging one to program in the rules of chemistry, clues to the identity of each atom given by the naming conventions of PDB, and robust exception handling. PDB parsing software often needs list of synonyms and tables of exceptions to correctly interpret the information. However this chapter is not intended to be a manual of how to construct a PDB parser.

Two newer chemical-based formats have emerged: mmCIF (MacroMolecular Chemical Interchange Format) and MMDB (Molecular Modeling Database Format). Both of these file formats are attempts to modernize PDB information. Both start by using data description languages, which are consistently machine parsable. The data description languages use "tag value" pairs, which are like variable names and values used in a programming language.

In both cases, the format specification is composed in a machine-readable form, and there is software that uses this format specification document to validate incoming streams of data. Both file formats are populated from PDB file data using the strategy of alignment-based reconstruction of the implicit ATOM and HETATM chemical graphs with the explicit SEQRES chemical graphs, together with extensive validation, which is recorded in the file. As a result, both of these file formats are superior for integrating with biomolecular sequence databases over PDB format data files, and their use in future software is encouraged.

mmCIF

The mmCIF file format was originally intended to be a biopolymer extension of the CIF (Chemical Interchange Format) familiar to small-molecule crystallographers and is based on a subset of the STAR syntax. CIF software for parsing and validating format specifications is not forward-compatible with mmCIF, since these have different implementations for the STAR syntax. The underlying data organization in an mmCIF record is a set of relational tables. The mmCIF project refers to their format specification as the mmCIF dictionary, kept on the Web at the Nucleic Acid Database site.

The mmCIF dictionary is a large document containing specification for holding the information stored in PDB files as well as many other data items derivable from the primary coordinate data, such as bond angles. The mmCIF data specification gives this data a consistent interface, which has been used to implement the NDB Protein Finder, a Web-based query format in a relational database style, and is also used as the basis for the new RCSB software systems.

Validating an incoming stream of data against the large mmCIF dictionary entails significant computational time: hence, mmCIF is probably destined to be an archival and advanced query format. Software libraries for reading mmCIF tables into relational tables into memory in FORTRAN and C are available.

MMDB

The MMDB file format is specified by means of the ASN.1 data description language, which is used in a variety of other settings, surprisingly enough including applications in telecommunications and automotive manufacturing. Because the US National Library of Medicine also uses ASN.1 data specifications for sequence and bibliographic information, the MMDB format borrows certain elements from other data specifications, such as the parts used in describing bibliographic references cited in the data record. ASN.1 files can appear as human-readable text files or as a variety of binary and packed binary files that can be decoded by any hardware platform. The MMDB standard residue dictionary is a lookup table of information about the chemical graphs of standard biopolymer residue types. The MMDB format specification is kept inside the NCBI toolkit distribution, but a browser is available over the Web for a quick look. The MMDB ASN.1 specification is much more compact and has fewer data items than the mmCIF dictionary, avoiding derivable data altogether.

In contrast to the relational table design of mmCIF, the MMDB data records are structured as hierarchical records. In terms of performance, ASN.1-formatted MMDB files provide for much faster input and output than do mmCIF or PDB records. Their nested hierarchy requires fewer validation steps at load time than the relational scheme in mmCIF or in the PDB file format; hence, ASN.1 files are ideal for three-dimensional structure database browsing.

A complete application programming interface is available for MMDB as part of the NCBI toolkit, containing a wide variety of C code libraries and applications. Both an ASN.1 input/output programming interface, layer and a molecular computing layer (MMDB-API) are present in the NCBI toolkit. The NCBI toolkit supports x86 and alpha-based Windows' platforms, Macintosh 68K and PowerPC' CPUs, and a wide variety of UNIX platforms. The three-dimensional structure database viewer (Cn3D) is an MMDB-API-based application with source code included in the NCBI toolkit.

Visualizing Structural Information

Multiple Representation Styles

We often use multiple styles of graphical representation to see different aspects of molecular structure. Here, the enzyme barnase 1BN1 appears both in wire-frame and space-filling model formats, as produced by RasMol.

Because the protein structure record 1BN1 has three barnase molecules in the crystallographic unit, the PDB file has been hand-edited using a text editor to delete the superfluous chains. Editing data files is an accepted and widespread practice in three-dimensional molecular structure software, forcing the three-dimensional structure viewer to show what the user wants. In this case, the crystallographic data recorded in the three-dimensional structure does not represent the functional ***biological*** unit. In our example, the molecule barnase is a monomer: however, we have three molecules in the crystallographic unit. In our other example, 3TS1, the molecule is a dimer, but only one of the symmetric subunits is recorded in the PDB file.

The wire-frame image clearly shows the chemistry of the barnase structure, and we can easily trace of the sequence of barnase on the image of its biopolymer in an interactive computer display. The space-filling model gives a good indication of the size and surface of the biopolymer, yet it is difficult to follow the details of chemistry and bonding in this representation. The composite illustration shows an α-carbon backbone in a typical pseudo-structure representation. The lines drawn are not actual chemical bonds, but they guide us along the path made by the α-carbons of the protein backbone. These are also called "virtual bonds." The purple tryptophan side chains have been selected and drawn together with a dot surface. This composite illustration highlights the volume taken up by the three tryptophan side chains in three hydrophobic core regions of barnase, while effectively hiding most of the structure's details.

The ribbon model shows the organization of the structural path of the secondary structure elements of the protein chain (α-helix and β-sheet regions). This representation is very often used, with the arrowheads indicating the N-to-C-terminal direction of the secondary structure elements, and is most effective for identifying secondary structures within complex topologies.

The variety of information conveyed by the different views illustrates the need to visualize three-dimensional biopolymer structure data in unique ways that are not common to other three-dimensional graphics applications. This requirement often precludes the effective use of software from the "macroscopic world," such as computer-aided design (CAD) or virtual reality modeling language (VRML) packages.

Picture the Data: Populations, Degeneracy, and Dynamics

Both X-ray and NMR techniques infer three-dimensional structure from a *synchronized* population of molecules—synchronized in space as an ordered crystal lattice or synchronized in behaviour as nuclear spin states are organized by an external magnetic field. In both cases, information is gathered from the population as a whole. The coordinate (x, y, z) locations of atoms in a structure are derived using numerical methods. These fit the expected chemical graph of the sample into the three-dimensional data derived from the experimental data. The expected chemical graph can include a mixture of biopolymer sequence-derived information as well as the chemical graph of any other known small molecules present in the sample, such as substrates, prosthetic groups, and ions.

One somewhat unexpected result of the use of molecular populations is the assignment of degenerate coordinates in a database record, i.e., more than one coordinate location for a single atom in the chemical graph. This is recorded when the population of molecules has observable conformational heterogeneity.

NMR Models and Ensembles

The NMR structures on the left appear "fuzzy." In fact, there are several different, complete structures piled one on top of another in these images. Each structure is referred to as a *model*, and the set of models is an *ensemble*. Each model in the ensemble is a chirally correct, plausible structure that fits the underlying NMR data as well as any other model in the ensemble.

The images from the ensemble of an NMR structure show the dynamic variation of a molecule in solution. This reflects the conditions of the experiment: molecules free in solution with freedom to pursue dynamic conformational changes. In contrast, the X-ray structures provide a very strong mental image of a static molecule. This also reflects the conditions of the experiment, an ordered crystal constrained in its freedom to explore its conformational dynamics. These mental images direct our interpretation of structure. If we measure distance between two atoms using an X-ray structure, we may get a single value. However, we can get a range of values for the same distance in each model looking at an ensemble of an NMR structure. Clearly, our interpretation of this distance can be dependent on the source of the three-dimensional structure data! It is prudent to steer clear of any software that ignores or fails to show the population degeneracy present in structure database records, since the absence of such information can further skew *biological* interpretations. Measuring the distance between two atoms in an NMR structure using software that hides the other members of the ensemble will give only one value and not the true range of distance observed by the experimentalist.

Correlated Disorder

Typically, X-ray structures have one and only one model. Some subsets of atoms, however, may have degenerate coordinates, which we will refer to as *correlated disorder.* Many X-ray structure database records show correlated disorder. Both correlated disorder and ensembles are often ignored by three-dimensional molecular graphics software. Some programs show only the first model in an ensemble, or the first location of each atom in a correlated disorder set, ignoring the rest of the degenerate coordinate values. Worse still, sometimes, erroneous bonds are drawn between the degenerate locations, making a mess of the structure.

Local Dynamics

A single technique can be used to constrain the conformation of some atoms differently from others in the same structure. For example, an internal atom or a backbone atom that is locked in by a multitude of interactions may appear largely invariant in NMR or X-ray data, whereas an atom on the surface of the molecule may have much more conformational freedom. Interior protein side chains typically show much less flexibility in ensembles, so it might be concluded that the interiors of proteins lack conformational dynamics altogether. However, a more sensitive, biophysical method, time-resolved fluorescence spectroscopy of single tryptophan residues, has a unique ability to detect heterogeneity (but not the actual coordinates) of the tryptophan side-chain conformation. Years of study using this method has shown that, time and time again, populations of interior tryptophans in pure proteins are more often in heterogeneous conformations than not. This method was shown to be able to detect rotamers of tryptophan within single crystals of erabutoxin, where X-ray crystallography could not. When interpreting three-dimensional structure data, remember that heterogeneity does persist in the data, and that the NMR and X-ray methods can be blind to all but the most populated conformations in the sample.

Database Structure Viewers

In the past several years, the software used to examine and display structure information has been greatly improved in terms of the quality of visualization and, more importantly, in terms of being able to relate sequence information to structure information.

Visualization Tools

Although the RCSB Web site provides a Java-based three-dimensional applet for visualizing PDB data, the applet does not currently support the display of nonprotein structures. For this and other reasons, the use of RasMol v2.7 is instead recommended for viewing structural data downloaded from RCSB; more information on RasMol appears in the following section. If a Java-based viewer is preferred, WebMol is recommended. With the advent of many homemade visualization programs that can easily be downloaded from the Internet, the reader is strongly cautioned to *only* use mature, well-established visualization tools that have been thoroughly tested and have undergone critical peer review.

RasMol and RasMol-Based Viewers

As mentioned above, several viewers for examining PDB files are available. The most popular one is RasMol. RasMol represents a breakthrough in software-driven three-dimensional graphics, and its source code is a recommended study material for anyone interested in high-performance three-dimensional graphics. RasMol treats PDB data with extreme caution and often recomputes information, making up for inconsistencies in the underlying database. It does not try to validate the chemical graph of sequences or structures encoded in PDB files. RasMol does not perform internally either dictionary-based standard residue validations or alignment of explicit and implicit sequences. RasMol 2.7.1 contains significant improvements that allow one to display information in correlated disorder ensembles and select different NMR models. It also is capable of reading mmCIF-formatted three-dimensional structure files and is thus the viewer of choice for such data. Other data elements encoded in PDB files, such as disulfide bonds, are recomputed based on rules of chemistry, rather than validated.

RasMol contains many excellent output formats and can be used with the Mol-script program to make wonderful PostScript™ ribbon diagrams for publication. To make optimal use of RasMol, however, one must master its command-line language, a familiar feature of many legacy three-dimensional structure programs.

Several new programs are becoming available and are free for academic users. Based on RasMol's software-driven three-dimensional-rendering algorithms and sparse PDB parser, these programs include Chime™, a Netscape™ plug-in. Another program, WebMol, is a Java-based three-dimensional structure viewer apparently based on RasMol-style rendering.

MMDB Viewer: Cn3D

Cn3D is a three-dimensional structure viewer used for viewing MMDB data records. Because the chemical graph ambiguities in data in PDB entries have been removed to make MMDB data records and because all the bonding information is explicit, Cn3D has the luxury of being able to display three-dimensional database structures consistently, without the parsing, validation, and exception-handling overhead required of programs that read PDB files. Cn3D's default image of a structure is more intelligently displayed because it works without fear of misrepresenting the data. However, Cn3D is dependent on the complete chemical graph information in the ASN.1 records of MMDB, and, as such, it does not read in PDB files.

Cn3D 3.0 has a much richer feature set than its predecessors, and it now allows selection of subsets of molecular structure and independent settings of rendering and colouring aspects of that feature. It has state-saving capabilities, making it possible to colour and render a structure, and then save the information right into the ASN.1 structure record, a departure from the hand-editing of PDB files or writing scripts. This information can be shared with other Cn3D users on different plateforms.

The images which are from Cn3D 3.0, now based on OpenGL three-dimensional graphics. This provides graphics for publication-quality images that are much better than previous versions, but the original Viewer3D version of Cn3D 3.0 is available for computers that are not capable of displaying OpenGL or that are too slow.

Also unique to Cn3D is a capacity to animate three-dimensional structures. Cn3D's animation controls resemble tape recorder controls and are used for displaying quickly the members of a multiple structure ensemble one after the other, making an animated three-dimensional movie. The GO button makes the images animated, and the user can rotate or zoom the structure while it is playing the animation. This is particularly useful for looking at NMR ensembles or a series of time steps of structures undergoing motions or protein folding. The animation feature also allows Cn3D to provide superior multiple structure alignment when used together with the VAST structure-structure comparison system, described later in this chapter.

Other 3D Viewers: Mage, CAD, and VRML

A variety of file formats have been used to present three-dimensional biomolecular structure data lacking in chemistry-specific data representations. These are viewed in generic three-dimensional data viewers such as those used for "macroscopic" data, like engineering software or virtual-reality browsers. File formats such as VRML contain three-dimensional graphical display information but little or no information about the underlying chemical graph of a molecule. Furthermore, it is difficult to encode the variety of rendering styles in such a file: one needs a separate VRML file for a space-filling model of a molecule, a wire-frame model, a ball-and-stick model, and so on, because each explicit list of graphics objects (cylinders, lines, spheres) must be contained in the file.

Biomolecular three-dimensional structure database records are currently not compatible with "macroscopic" software tools such as those based on CAD software. Computer-aided design software represents a mature, robust technology, generally superior to the available molecular structure software. However. CAD software and file formats in general are ill-suited to examine the molecular world, owing to the lack of certain "specialty" views and analytical functions built in for the examination of details of protein structures.

Making Presentation Graphics

To get the best possible publication-quality out of any molecular graphics software, first consider whether a bitmap or a vector-based graphic image is needed. Bitmaps are made by programs like RasMol and Cn3D—they reproduce exactly what you see on the screen, and are usually the source of trouble in terms of pixellation ("the jaggies"), a bitmap of 380-400 pixels. High-quality print resolution is usually at 300-600 dots per inch, but monitors have far less information in pixels per inch (normally 72 dpi), so a big image on a screen is quite tiny when printed at the same resolution on a printer. Expanding the image to fit a page causes exaggeration of pixel steps on diagonal lines.

The best advice for bitmaps is to use as big a monitor/desktop as possible, maximizing the number of pixels included in the image. This may mean borrowing a colleagues' 21-in monitor or using a graphics card that offers a "virtual desktop" that is larger than the monitor being used in pixel count. In any case, always fill the entire screen with the viewer window before saving a bitmap image for publication.

Advanced Structure Modeling

Tools that go beyond simple visualization are now emerging and are freely available. Biologists often want to display structures with information about charge distribution, surface accessibility, and molecular shape; they also want to be able to perform simple mutagenesis experiments and more complex structure modeling. SwissPDB Viewer also known as Deep View, is provided free of charge to academics and can address a good number of these needs. It is a multi platform (Mac, Win, and Linux) OpenGL-based tool that has the ability to generate molecular surfaces, align multiple proteins, use scoring functions, as well as do simple, fast modeling, including site-directed mutagenesis and more complex modeling such as loop rebuilding. An excellent tutorial for SwissPDB Viewer developed by Gale Rhodes is one of the best starting points for making the best use of this tool. It has the capability to dump formatted files for the free ray-tracing software POV-Ray, and it can be used to make stunning images of molecular structures, easily suitable for a journal cover.

Structural Similarity Searching

Although a sequence-sequence similarity program provides an alignment of two sequences, a structure-structure similarity program provides a three-dimensional structural superposition. This superposition results from a set of three-dimensional rotation-translation matrix operations that superimpose similar parts of the structure. A conventional sequence alignment can be derived from three-dimensional superposition by finding the α-carbons in the protein backbone that are superimposed in space. Structure similarity search services are based on the premise that some similarity metric can be computed between two structures and used to assess their similarity, much in the same way a BLAST alignment is scored. A structure similarity search service can take a three-dimensional protein structure, either already in PDB or a new one, and compare that structure, making three-dimensional superpositions with other structures in the database and reporting the best match without knowing anything about the sequence. If a match is made between two structures that are not related by any measurable sequence similarity, it is indeed a surprising discovery. For this type of data to be useful, the similarity metric *must* be meaningful. A large fraction of structures, for example, have β-sheets. Although a similar substructure may include a single β-hairpin turn with two strands, one can find an incredibly large number of such similarities in the PDB database, so these similarities are simply not surprising or informative. A number of structure similarity searching systems are now available on the Internet, and almost all of them can be found following links from the RCSB Structure Summary page. The process of similarity searching presents some interesting high-performance computational challenges, and this is addressed in different ways, ranging from human curation, as the SCOP system provides, to fully automated systems, such as DALI, SCOP, or the CE system provided by RCSB.

The Vector Alignment Search Tool (VAST) provides a similarity measure of three-dimensional structure. It uses vectors derived from secondary structure elements, with no sequence information being used in the search. VAST is capable of finding structural similarities when no sequence similarity is detected. VAST, like BLAST, is run on all entries in the database in an $N \times N$ manner, and the results are stored for fast retrieval using the Entrez interface. More than 20,000 domain substructures within the current three-dimensional structure database have been compared with one another using the VAST algorithm, the structure-structure superpositions recorded, and alignments of sequence derived from the superposition. The VAST algorithm focuses on similarities that are surprising in the *statistical* sense. One does not waste time examining many similarities of small substructures that occur by chance in protein structure comparison. For example, very many small segments of β-sheets have obvious, but not surprising, similarities. The similarities detected by VAST are often examples of remote homology, undetectable by sequence comparison. As such, they may provide a broader view of the structure, function, and evolution of a protein family.

The VAST system stands out amongst these comparative tools because (a) it has a clearly defined similarity metric leading to surprising relationships, (b) it has an adjustable interface that shows nonredundant hits for a quick first look at the most interesting relationships, without seeing the same relationships lots of times, (c) it provides a domain-based structure comparison rather than a whole protein comparison, and (d) it has the capability to integrate with Cn3D as a visualization tool for inspecting surprising structure relationships in detail. In addition to a listing of similar structures, VAST-derived structure neighbours contain detailed residue-by-residue alignments and three-dimensional transformation matrices for structural superposition, in practice, refined alignments from VAST appear conservative, choosing a highly similar "core" substructure compared

Table 8.1. Important internet resources.

BIND	*http://bioinfo.mshri.on.ca*
Imagemagick	*http://http://www.wizards.dupont.com/cristy/ImageMagick.html*
mmCif Project	*http://ndbserver.rutgers.edu/NDB/mmcif/index.html*
National Center for Biotechnology Information (NCBI)	*http://www.ncbi.nlm.nih.gov/*
NCBI Toolkit	*http://www.ncbi.nlm.nih.gov/Toolbox*
Nucleic Acid Database (NDB)	*http://ndbserver.rutgers.edu/*
POV-RayAY	*http://www.povrays.org/*
Protein Data Bank at RCSB	*http://www.rcsb.org/*
RasMol	*http://www.bernstein-plus-sons.com/*
SwissPDB Viewer/Deep view	*http://http://www.expasy.ch/spdbv/mainpage.html*
WebMol	*http://www.cmpharm.ucsf.edu/~walther/webmol/*

with DALI superpositions. With the VAST superposition, one easily identifies regions in which protein evolution has modified the structure, whereas DALI superpositions may be more useful for comparisons involved in making structural models. Both VAST and DALI superpositions are excellent tools for investigating relationships in protein structure, especially when used together with the SCOP database of protein families.

9

Phylogenetic Analysis

Phylogenetics is the study of evolutionary relationships. Phylogenetic analysis is the means of *inferring* or estimating these relationships. The evolutionary history inferred from phylogenetic analysis is usually depicted as branching, treelike diagrams that represent an estimated pedigree of the inherited relationships among molecules ("gene trees"), organisms, or both. Phylogenetics is sometimes called cladistics because the word "clade," a set of descendants from a single ancestor, is derived from the Greek word for branch. However, cladistics is a particular method of hypothesizing about evolutionary relationships.

The basic tenet behind cladistics is that members of a group or clade share a common evolutionary history and are more related to each other than to members of another group. A given group is recognized by sharing unique features that were not present in distant ancestors. These shared, *derived* characteristics can be anything that can be observed and described—from two organisms having developed a spine to two sequences having developed a mutation at a certain base pair of a gene. Usually, cladistic analysis is performed by comparing multiple characteristics or "characters" at once, either multiple phenotypic characters or multiple base pairs or amino acids in a sequence.

1. There are three basic assumptions in cladistics: Any group of organisms is related by descent from a common ancestor (fundamental tenet of evolutionary theory).
2. There is a bifurcating pattern of cladogenesis. This assumption is controversial.
3. Change in characteristics occurs in lineages over time. This is a necessary condition for cladistics to work.

The resulting relationships from cladistic analysis are most commonly represented by a phylogenetic tree:

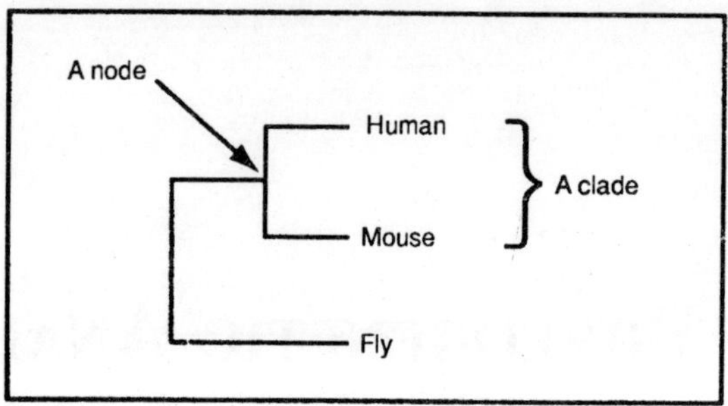

Even with this simple tree, a number of terms that are used frequently in phylogenetic analysis can be introduced:

1. A *clade* is a monophyletic taxon. Clades are groups of organisms or genes that include the most recent common ancestor of all of its members and all of the descendants of that most recent common ancestor. Clade is derived from the Greek word "klados," meaning branch or twig.
2. A *taxon* is any named group of organisms but not necessarily a clade.
3. In some analyses, *branch* lengths correspond to divergence (e.g., in the above example, mouse is slightly more related to fly than human is to fly).
4. A *node* is a bifurcating branch point.

Macromolecules, especially sequences, have surpassed morphological and other organismal characters as the most popular form of data for phylogenetic or cladistic analysis. It is this molecular phylogenetic analysis that we will introduce here.

It is unrealistic to believe that an all-purpose phylogenetic analysis recipe can be delineated. Although numerous phylogenetic algorithms, procedures, and computer programs have been devised, their reliability and practicality are, in all cases, dependent on the structure and size of the data. The merits and pitfalls of various methods are the subject of often acrimonious debates in taxonomic and phylogenetic journals. Some of these debates are summarized in a series of useful reviews of phylogenetics. An especially concise introduction to molecular phylogenetics is provided by Hillis et al. (1993).

The danger of generating incorrect results is inherently greater in computational phylogenetics than in many other fields of science. The events yielding a phylogeny happened in the past and can only be inferred or estimated. Despite the well-documented limitations of available phylogenetic procedures, current biological literature is replete with examples of conclusions derived from the results of analyses in which data had been simply run through one or another phylogeny program. Occasionally, the limiting factor in phylogenetic analysis is not so much the computational method used; more often than not, the limiting factor is the users' understanding of what the method is actually doing with the data. This brief guide to phylogenetic analysis has several objectives. First, a conceptual approach that describes some of the most important principles underlying the most widely and easily applied methods of phylogenetic analyses of biological sequences and their interpretation will be introduced. The aim is to show that practical phylogenetic analysis should be conceived as a search for a correct model, as much as a search for the correct tree. In this context, some of the particular models assumed by various popular methods and how these models might affect analysis of particular data sets will be discussed. Finally, some examples of the application of particular methods to the inferences of evolutionary history are provided.

Note that the principles for DNA analysis will be initially discussed, although most also apply to protein sequences (except where further description of protein sequences is indicated). As there is a growing interest in the analysis of protein sequences, the reader is directed to further descriptions of protein-specific problems.

Fundamental Elements of Phylogenetic Models

Phylogenetic tree-building methods presume particular evolutionary models. For a given data set, these models can be violated because of occurrences such as the transfer of genetic material between organisms. Thus, when interpreting a given analysis, one should always consider the model used and its assumptions and entertain other possible explanations for the observed results. An investigation of organismal relationships in the tree suggests the eukaryote 1 is more related to the bacteria than to the other eukaryotes. Because the vast majority of other cladistic analyses, including those based on morphological features, suggest that eukaryote 1 is more related to the other eukaryotes than to bacteria, we suspect that for this analysis the assumptions of a bifurcating pattern of evolution are incorrect. We suspect that horizontal gene transfer from an ancestor of the bacteria 1, 2, and 3 to the ancestor of eukaryote 1 occurred because this would most simply explain the results.

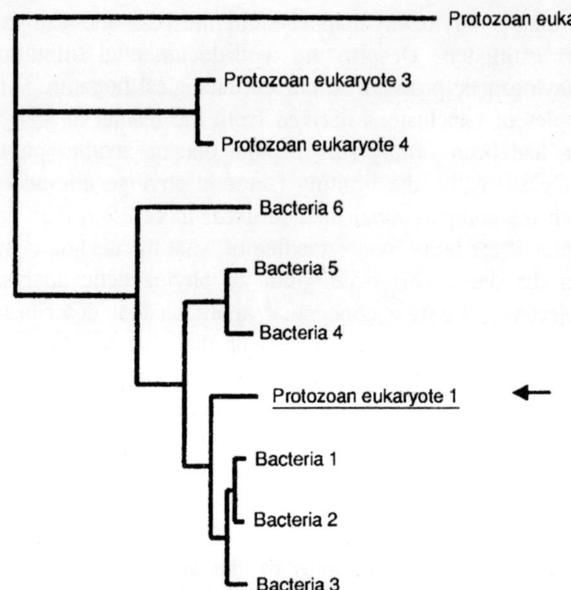

Fig. 9.1. Example of a phylogenetic tree based on genes that do not match organismal phylogeny, suggesting horizontal gene transfer has occurred.

Models inherent in phylogenetics methods make additional "default" assumptions:

1. The sequence is correct and originates from the specified source.
2. The sequences are homologous (i.e., are all descended in some way from a shared ancestral sequence).
3. Each position in a sequence alignment is homologous with every other in that alignment.
4. Each of the multiple sequences included in a common analysis has a common phylogenetic history with the others (e.g., there are no mixtures of nuclear and organellar sequences).
5. The sampling of taxa is adequate to resolve the problem of interest.
6. Sequence variation among the samples is representative of the broader group of interest.
7. The sequence variability in the sample contains phylogenetic signal adequate to resolve the problem of interest.

There are additional assumptions that are defaults in some methods but can be at least partially corrected for in others:

1. The sequences in the sample evolved according to a single stochastic process.

2. All positions in the sequence evolved according to the same stochastic process.
3. Each position in the sequence evolved independently.

Errors in published phylogenetic analyses can often be attributed to violations of one or more of the foregoing assumptions. Every sequence data set must be evaluated against these assumptions, with other possible explanations for the observed results considered.

Tree Interpretation—Importance of Identifying Paralogs and Orthologs

As more genomes are sequenced, we are becoming more interested in learning about protein or gene evolution (i.e., investigating gene phylogeny, rather than organismal phylogeny). This can aid our understanding of the function of proteins and genes.

Studies of protein and gene evolution involve the comparison of *homologs*—sequences that have common origins but may or may not have common activity. Sequences that share an arbitrary, threshold level of similarity determined by alignment of matching bases are termed *homologous*. They are inherited from a common ancestor that possessed similar structure, although the structure of the ancestor may be difficult to determine because it has been modified through descent.

Homologs are most commonly either orthologs, paralogs, or xenologs.

Orthologs are homologs produced by speciation. They represent genes derived from a common ancestor that diverged due to divergence of the organisms they are associated with. *They tend to have similar function*.

Paralogs are homologs produced by gene duplication. They represent genes derived from a common ancestral gene that duplicated within an organism and then subsequently diverged. *They tend to have different functions*.

Xenologs are homologs resulting from horizontal gene transfer between two organisms. The determination of whether a gene of interest was recently transferred into the current host by horizontal gene transfer is often difficult. Occasionally, the %(G + C) content may be so vastly different from the average gene in the current host that a conclusion of external origin is nearly inescapable, however often it is unclear whether a gene has horizontal origins. Function of xenologs can be variable depending on how significant the change in context was for the horizontally moving gene; however, in general, the function tends to be similar.

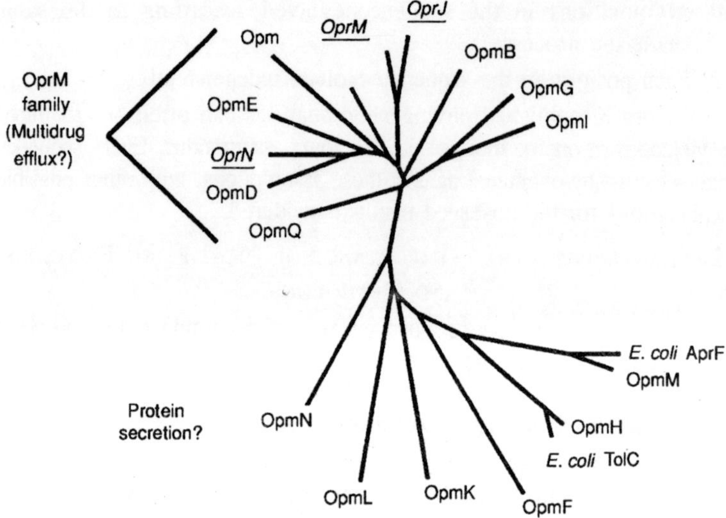

Fig. 9.2. Insight into protein function from an investigation of paralogs and orthologs—an example.

Phylogenetic Data Analysis

A straightforward phylogenetic analysis consists of four steps:

1. Alignment (both building the data model and extracting a phylogenetic dataset)
2. Determining the substitution model
3. Tree building
4. Tree evaluation

Each step is critical for the analysis and should be handled accordingly. For example, trees are only as good as the alignment they are based on. When performing a phylogenetic analysis, it often insightful to build trees based on different modifications of the alignment to see how the alignment proposed influences the resulting tree.

Alignment: Building Data Model

Phylogenetic sequence data usually consist of multiple sequence alignments; the individual, aligned-base positions are commonly referred to as "sites." These sites are equivalent to "characters" in theoretical phylogenetic discussions, and the actual base (or gap) occupying a site is the "character state."

This chapter reviews similar alignment methods in the context of phylogenetic analysis. Aligned sequence positions subjected to phylogenetic analysis represent a priori phylogenetic conclusions because the sites themselves (not the actual bases) are effectively assumed to be genealogically related, or homologous. Sites at which one is confident of homology and that contain changes in character states useful for the given phylogenetic analysis are often referred to as "informative sites."

Steps in building the alignment include selection of the alignment procedure(s) and extraction of a phylogenetic data set from the alignment. The latter procedure requires determination of how ambiguously aligned regions and insertion/deletions (referred to as *indels*, or gaps) will be treated in the tree-building procedure.

A typical alignment procedure involves the application of a program such as CLUSTAL W, followed by manual alignment editing and submission to a tree-building program. This procedure should be performed with the following questions and considerations in mind:

- *How much computer dependence*? Fully computational multiple alignment is sometimes advocated on the grounds that manual editing is inexplicit and/or unobjective. Usually, however, manual alignment editing is advocated because alignment algorithms and programs are not optimally adapted for phylogenetic alignment.
- *Phylogenetic criteria preferred*. Some computational multiple alignment methods align sequences strictly based on the order they receive them (the input order) without any consideration of their relationship. However, many current methods (e.g., CLUSTAL W, PileUp, ALIGN in ProPack) align according to an explicitly phylogenetic criterion (a "guide tree"). These guide trees are generated on the basis of initial pairwise sequence alignments. SAM and MACAW are examples of multiple alignment programs that do not explicitly invoke phylogenetic criteria, although it is possible to manipulate parameters in these programs to mimic phylogenetic processes. Theory holds that more closely related sequences should be aligned first and then the resulting groups of sequences, which may be less related to one another but still have a common ancestor, should share the same ancestral indels. This means that they could then be more accurately aligned.

The guide tree from CLUSTAL W is formatted as a PHYLIP tree file and can be imported in various tree-drawing programs. Some programs are designed to simultaneously (recursively) optimize an alignment and a phylogenetic tree (e.g., TreeAlign and MALIGN). In theory, an optimal simultaneous solution or set of solutions to an alignment/phylogeny problem exists, but the hazard of the recursive approach lies in the possibility of funneling the analysis toward a

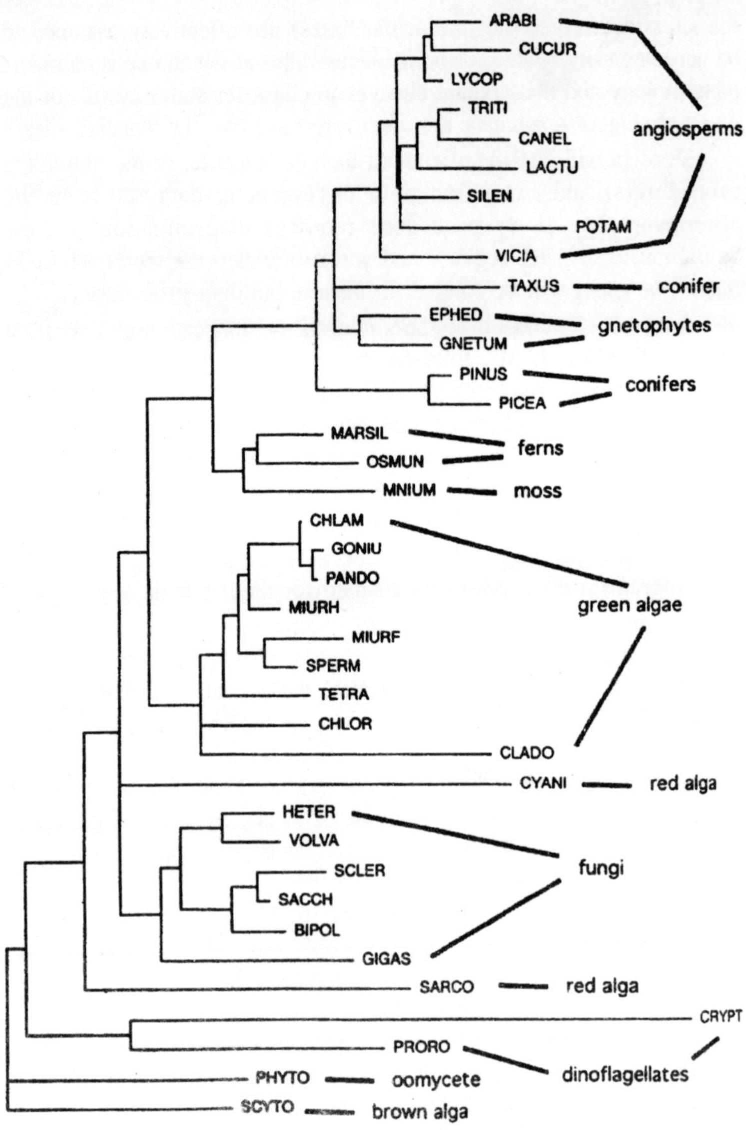

Fig. 9.3. CLUSTAL guide tree for 5.8S rDNA sequences of selected plants, fungi, and protists.

wrong or incomplete solution. Thus, when the tree-building analysis based on the alignment is followed, one should consider whether other evolutionary relationships might be favoured using a slightly modified alignment.

- *Alignment parameter estimation.* The most important parameters in an alignment method are those that determine the placement of indels or gaps in an alignment of length-variable sequences. Alignment parameters should vary dynamically with evolutionary divergence, such that base mismatches are more likely as the sequences become more divergent. Alignment parameters should also be adjusted to prevent closely related, overrepresented sequences from adversely influencing the alignment of under-represented sequences. This is accomplished by downweighting the alignment score contribution of closely related sequences. These dynamic parameter adjustments are both implemented in CLUSTAL W, whereas sequence weighting is implemented in SAM.
- *Which alignment procedure is best for phylogenetic analysis?* The short answer is "the method that is closest to understanding the evolutionary relationships between the sequences being examined." Unless the actual phylogenetic relationships are known beforehand, there is no clear way to determine which alignment procedure is best for a given phylogenetic analysis. In general, it is inadvisable to simply subject a computer-generated alignment to a tree-building procedure because the latter is blind to errors in the former. This caution especially applies to tree-building programs included in alignment packages (e.g., CLUSTAL W and TREE in ProPack) because the tree-building methods in these programs are not rigorous. However, as long as the entire alignment is scrutinized in view of independent phylogenetic evidence, methods such as CLUSTAL W that utilize some degree of phylogenetic criteria are among the best currently available.
- *Mathematical optimization and analysis of structures.* Some alignment programs (e.g., MACAW, SAM) optimize according to a statistical model, but the relationship of these statistics to phylogenetic models is not yet clear. No methods are yet available for determining whether one multiple alignment is significantly better than another based on a phylogenetic model.

Aligning according to secondary or tertiary sequence structure is considered phylogenetically more reliable than sequence-based alignment because confidence in homology assessment is greater when comparisons are made to complex structures rather than to simple characters (primary sequence). However, there does not appear to be any way to computationally facilitate phylogeny-based structural multiple alignment. Hopefully, new insights into these areas will be developed in the near future.

ALIGNMENT: EXTRACTION OF A PHYLOGENETIC DATA SET

In alignments that include length variation, the phylogenetic data set is usually not identical to the alignment. Even in alignments of length-invariable sequences, the data set can be different—for example, when only first and second codon positions are to be analyzed to avoid the strong G + C bias in the third codon position from affecting the final results.

In the case of length-variable sequences, the degree of difference between an alignment and the phylogenetic data set is determined mainly by how alignment ambiguities and indels are treated. The most extreme way to treat indels is to remove from the analysis all sites that include gaps. This approach has the advantage of permitting all the variation in the sequences to be described in terms of the substitution model, without the need for an ad hoc model to account for indels. The disadvantage of this approach is that phylogenetic signals contained within the indel regions are discarded.

Maximum parsimony is the only method that permits for the incorporation of alignable gaps as characters. These can be included in either of two ways: as an additional character state (a "fifth" nucleotide base or "twenty-first" amino acid) or as a set of characters independent of base substitutions. The first approach is not tenable for gaps occupying more than one site, for these will be counted as independent character state changes. The latter approach is useful for analyzing an alignment in which subsets of sequences contain perfectly aligned gaps. A set of gap characters can be appended to the aligned sequence data set, or the gaps can be scored "in place" by using the extra base approach but scoring only one of the gap positions in a sequence as a gap and the remainder as missing. These approaches can be implemented using PAUP.

For some alignments, procedures that ignore all gap scores or all sites including gap scores are less than ideal. However, there is not yet any program that allows one to ignore individual sites in individual sequences. When alignment might be unambiguous within groups of sequences but ambiguous among them, alignment "surgery" is warranted to ensure that unambiguous information relevant to groups of sequences can be retained and ambiguous information removed.

In gapped regions, one should determine whether alternative alignments seem reasonably plausible and, just as important, whether they might bias the tree-building analysis. When alignment ambiguities are resolved manually, phylogenetic relations, substitution processes, and base composition should be considered. It is perfectly reasonable at this stage to resolve ambiguities in favour of phylogenetic evidence and in some cases to delete ambiguous regions in the alignment. The advantage of this latter approach is that unambiguous information relevant to particular sequences can be retained over ambiguous data. The disadvantage is that parsimony and likelihood tree-building methods can interpret the "missing" information as zero divergence.

In summary, the following points should be considered when constructing a multiple sequence alignment for a phylogenetic analysis:

1. The alignment step in phylogenetic analysis is one of the most important because it produces the data set on which models of evolution are used.
2. It is not uncommon to edit the alignment, deleting unambiguously aligned regions and inserting or deleting gaps to more accurately reflect probable evolutionary processes that led to the divergence between sequences.
3. It is useful to perform phylogenetic analyses based on a series of slightly modified alignments to determine how ambiguous regions in the alignment affect the results and what aspects of the results one may have more or less confidence in.

Determining the Substitution Model

The substitution model should be given the same emphasis as alignment and tree building. As implied in the preceding section, the substitution model influences both alignment and tree building; hence, a recursive approach is warranted. At the present time, two elements of the substitution model can be computationally assessed for nucleotide data but not for amino acid or codon data. One element is the model of substitution between particular bases; the other is the relative rate of overall substitution among different sites in the sequence. Simple computational procedures have not been developed for assessing more complex variables (e.g., site-or lineage-specific substitution models). An overview of substitution models is presented below.

Models of Substitution Rates Between Bases

In general, substitutions are more frequent between bases that are biochemically more similar. In the case of DNA, the four types of transition (A $\rightarrow$ G, G $\rightarrow$ A, C $\rightarrow$ T, T $\rightarrow$ C) are usually more frequent than the eight types of transversion (A $\rightarrow$ C, A $\rightarrow$ T, C $\rightarrow$ G, G $\rightarrow$ T, and the reverse). Such biases will affect the estimated divergence between two sequences.

Specification of the relative rates of substitution among particular residues usually takes the form of a square matrix; the number of rows/columns is four in the case of bases, 20 in the case of amino acids (e.g., in PAM and BLOSUM matrices) and 61 in the case of codons (excluding stop codons). The off-diagonal element of the matrix correspond to the relative costs of going from one base to another. The diagonal elements represent the cost of having the same base in different sequences.

The cost schedule can be fixed a priori to ensure that the tree-building method will tally an exact cost for each substitution incurred. Fixed-cost matrices are character-state weight matrices and are applied in maximum parsimony (MP) tree building. When such weights are applied, the method is referred to as "weighted parsimony." For

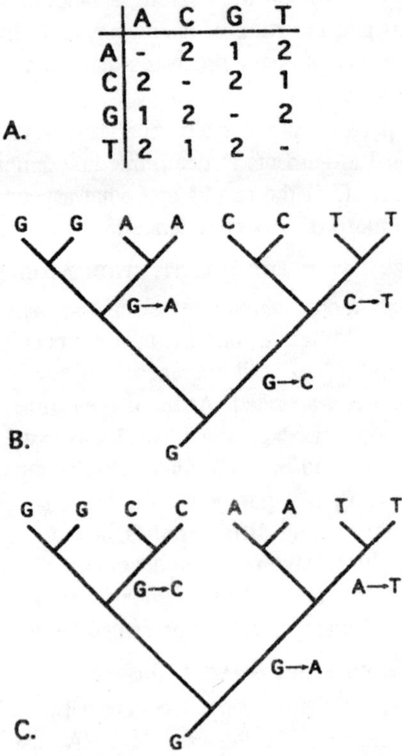

Fig. 9.4. Character weight matrix and application in MP phylogenetic analysis. (A) Matrix indicating that a transverson substitution coast twice that of a transition. (B, C) Two phylogenetic resolutions and reconstructions of the evolution of a hypothetical pattern of aligned bases at a particular site in eight sequences.

distance and maximum likelihood (ML) tree building, the costs can be derived from instantaneous rate matrices representing ML estimators of the probability that a particular type of substitution will occur. Although application of the MP weight matrix is just simple arithmetic, application of the distance and ML rate matrices can involve complex algebra. To avoid the blind application of possibly inappropriate methods, practitioners are advised to familiarize themselves with the relevant underlying theory.

Character-state weight matrices have usually been estimated more or less by eye, but they can also be derived from a rate matrix. For example, if it is presumed that each of the two transitions occurs at double the frequency of each transversion, a weight matrix can simply specify, for example, that the cost of A-G is 1 and the cost of A-T is 2. (The parsimony method dictates that the diagonal elements of the matrix, or the cost of having the same base in different sequences, be zero. This proves to be a shortcoming of parsimony; this will be discussed further below.) In the subsequent tree-building step, this set of assumptions will minimize the overall number of transversions and tend to cluster sequences differing mainly by transitions.

A simplified substitution rate matrix used in ML and distance phylogenetic analysis is presented in Figure 9.5. The matrix is analogous to that presented in Figure 9.4, but the actual computation of divergence involves more complex algebra and cannot be determined by simply counting steps between bases.

The paralinear or "log-det" transformation corrects for nonstationarity. In this method, which is applicable only to distance tree building, the numbers of raw substitutions of each type and in each direction are tallied for each sequence pair in a four-by-four matrix. Each matrix has an algebraic determinant, the log of which becomes a factor in estimating sequence divergence, hence the name "log-det." Pairwise comparisons of sequences having various and assorted patterns of base frequencies will yield a variety of matrix patterns, giving a variety of determinant values. Thus, each estimated

$$\begin{array}{c} \\ A \\ C \\ G \\ T \end{array} \begin{array}{c} \begin{array}{cccc} A & C & G & T \end{array} \\ \begin{bmatrix} -(a_1+a_2+a_3) & a_1 & a_2 & a_3 \\ a_4 & -(a_4+a_5+a_6) & a_5 & a_6 \\ a_7 & a_8 & -(a_7+a_8+a_9) & a_9 \\ a_{10} & a_{11} & a_{12} & -(a_{10}+a_{11}+a_{12}) \end{bmatrix} \end{array}$$

Fig. 9.5. Simplified substitution rate matrix used in ML and distance phylogenetic analysis.

		Sclerotinium sclerotiorum				
		A	C	G	T	total
Spinacia oleracea	A	340	6	13	4	363
	C	10	229	6	36	281
	G	25	8	229	12	372
	T	5	22	6	312	345
	total	380	265	352	364	1361

Fig. 9.6. Pairwise sequence comparison.

pairwise distance will be affected by the determinant particular to each pair, which effectively allows the substitution model to be different for each, varying along different branches of a phylogenetic tree. Log-det is especially sensitive to among-site rate heterogeneity, since base frequency bias can exist only in sites that are subject to variation.

Models of Among-Site Substitution Rate Heterogeneity

In addition to variation in substitution patterns, variation in substitution rates among different sites in a sequence has been shown to profoundly affect the results of tree building. The most obvious example of among-site rate variation, or heterogeneity, is that evident among the three codon positions in a coding sequence. Due to the degeneracy of the genetic code, changes in the third codon position can more frequently occur without affecting the ultimately encoded protein sequence. Therefore, this third codon position tends to be much more variable than the first two. For this reason, many phylogenetic analyses of coding sequences exclude the third codon position. In some cases, however, rate variation patterns are more subtle (e.g., those corresponding to conserved regions of proteins or rRNA).

Approaches to the estimation of substitution rate heterogeneity are the nonparametric models, the invariants model, and the gamma distribution models. The nonparametric approach derives categories of relative rates for particular sites. This approach can be used with MP tree building simply by weighting particular sites according to relative mutation frequency, although such weighting tends to require prior knowledge of the true tree. The approach is also applicable to ML tree building, but it is considered computationally impractical. The invariants approach estimates a proportion of sites that are not free to vary. The remaining sites are presumed to vary with equal probability. The gamma approach assigns a substitution probability to sites by assuming that, for a given sequence, the probabilities vary according to a gamma distribution. The shape of the gamma distribution, as described by the shape parameter, describes the distribution of substitution probabilities among sites in a sequence. In a combined approach, it can be presumed that a proportion of sites are invariant and that the remainder varies according to a gamma distribution.

In practice, gamma correction can be continuous, discrete, or "autodiscrete". "Continuous gamma" means that sites are assigned to a change probability along a continuous curve. At present, this approach is computationally impractical in most cases. The discrete gamma approximation assigns sites to a specified number of categories that approximate the shape of the gamma curve. The autodiscrete model assumes that adjacent sites have correlated rates of change. Groups of sites are assigned to categories, and sites within a category can be assumed to have either constant or heterogeneous rates.

Various rate heterogeneity corrections are implemented in several tree-building programs. For nucleotide data, PAUP 4.0 implements both invariants and discrete gamma models for separate or combined use with time-reversible distance and likelihood tree-building methods and invariants in conjunction with the log-det distance method. For nucleotide, amino acid, and codon data, PAML implements continuous, discrete, and autodiscrete models. For nucleotide and amino acid data, PHYLIP implements a discrete gamma model.

Models of Substitution Rates between Amino Acids

The most widely used models of amino acid substitution include distance-based methods, which are based on matrixes such as PAM and BLOSUM. Briefly, Dayhoff's PAM 001 matrix is an empirical model that scales probabilities of change from one amino acid to another in terms of an expected 1% change between two amino acid sequences. This matrix is used to make a transition probability matrix that allows prediction of the probability of changing from one amino acid to another and also predicts equilibrium amino acid composition. Phylogenetic distances are calculated with the assumption that the probabilities in the matrix are correct. The distance that is computed is scaled in units of expected fraction of amino acids changed. Kimura's distance is another method used in PROTDIST, one of the PHYLIP family of programs, and is a rough distance formula for approximating PAM distance by simply measuring the fraction of amino acids that differ between two sequences and computing the distance by a set formula. This is a more rapid method, but it has some obvious limitations. It does not take into account which amino acids differ or what amino acids are changed, so some information is lost. The distance measure is represented as the fraction of amino acids differing; this is also the case with PAM distances. If the fraction of amino acids differing gets larger than 0.8541, the distance becomes infinite. Although PROTDIST is one of the most widely used programs providing substitution models for calculating protein distances, others that are faster and make use of additional matrices such as BLOSUM are now more widely-used (e.g., PUZZLE).

The model used in parsimony (not a distance-based method) insists that any amino acid changes be consistent with the genetic code so that, for example, lysine is allowed to substitute to methionine but not to proline. However, changes between two amino acids via a third are allowed and are counted as two changes if each of the two replacements is individually allowed. This sometimes allows changes that, at first sight, one would think should be outlawed. Thus, phenylalanine can be changed to glutamine via leucine in two steps total. Genetic code translation tables show that there is a leucine codon one step away from a phenylalanine codon; there is also a leucine codon one step away from a glutamine codon. These leucine codons, however, are not identical. It actually takes three base substitutions to get from either of the phenylalanine codons (UUU and UUC) to either of the glutamine codons (CAA or CAG). Why, then, does this program count only two? The answer is that recent DNA sequence comparisons seem to show that synonymous changes (changes in the nucleotide sequence of a codon region that do not change what amino acids are encoded by that region) are considerably faster and easier than ones that change the amino acid outright. We are assuming that, in effect, synonymous changes occur so much more readily that they need not be counted. Thus, in the chain of changes UUU (Phe) → CUU (Leu) → CUA (Leu) → CAA (Glu), the middle one is not counted because it does not actually change the amino acid (leucine).

Which Substitution Model to Use?

Although any of the parameters in a substitution model might prove critical for a given data set, the best model is not always the one with the most parameters. To the contrary, the fewer the parameters, the better. This is because every parameter estimate has an associated variance. As additional parametric dimensions are introduced, the overall variance increases, sometimes prohibitively. For a given DNA sequence comparison, a two-parameter model will require that the summed base differences be sorted into two categories and into six for a six-parameter model. Obviously, the number of sites sampled in each of the six categories would be much smaller (and perhaps too small) to give a reliable estimate.

A good strategy for substitution model specification for DNA sequences is the "describe tree" feature in PAUP, which uses likelihood to simultaneously estimate the six reversible substitution rates, the shape parameter of the gamma distribution, and the proportion of invariant sites. These parameters can be estimated by means of equal or specified base frequencies. Usually, any reasonable phylogenetic tree (e.g., an easily generated neighbour-joining tree) is suitable for this procedure because parameter estimates are apparently influenced predominantly by the character pattern rather than by the tree topology. This estimation procedure is not overly time consuming for up to 50 sequences. If there will be more sequences or less time, the test tree can be selectively pruned to reduce the number of taxa while retaining the overall phylogenetic range and structure. From the estimated substitution parameters, one can determine whether a simpler model is justified (e.g., whether the six substitution categories can be reduced to two) by comparing likelihood scores estimated for this tree using more or fewer parameters. Parameters for and the proportion of invariant sites sometimes can substitute for each other, so one should compare likelihoods with each estimated alone versus both together. Note that, unlike MP and ME, the ML scores derived using different parameter values are directly comparable.

In the case of protein-coding DNA sequences, it is sometimes obvious that, depending on the divergence of the samples, the useful variation is essentially either in the first and second codon positions, with the third positions randomized across the data set, or in the third position, with the first and second positions invariant. The procedure above will correct for this rate heterogeneity, although removing the "useless" sites may permit a more precise estimate of rate heterogeneity in the remaining sites.

For protein sequences, the model used is often dependent on the degree of sequence similarity. For more divergent sequences, the BLOSUM matrices are often better, whereas the PAM matrix is suited for more highly similar sequences. Both parsimony and distance matrix methods have benefits and disadvantages, and their use depends on one's philosophy about protein sequence changes: Is it better to retain information about each character when determining a tree (i.e., through parsimony) or to derive distance measures to base the tree (i.e., using a distance matrix)? Is a matrix based on empirical data a more accurate reflection of evolutionary change than a matrix based on generated theories about sequence change? Again, although cladistic analysis can be a powerful method for investigating evolutionary relationships, keep in mind that there is no one clear method that is better than the other. Each has its own benefits and disadvantages that differ depending on the type of analyses performed and the philosophy of the investigator.

Tree-Building Methods

Tree-building methods implemented in available software are discussed in detail in the literature and described on the Internet. This section briefly describes some of the most popular methods. Tree-building methods can be sorted into distance-based vs. character-based methods. Much of the discussion in molecular phylogenetics dwells on the utility of distance- and character-based methods. Distance methods compute pairwise distances according to some measure and then discard the actual data, using only the fixed distances to derive trees. Character-based methods derive trees that optimize the distribution of the actual data patterns for each character. Pairwise distances are, therefore, not fixed, as they are determined by the tree topology. The most commonly applied distance-based methods include neighbour-joining and the Fitch-Margoliash method, and the most common character-based methods include maximum parsimony and maximum likelihood.

Distance-Based Methods

Distance-based methods use the amount of dissimilarity (the distance) between two aligned sequences to derive trees. A distance method would reconstruct the true tree if all genetic divergence events were accurately recorded in the sequence. However, divergence encounters an upper limit as sequences become mutationally saturated. After one sequence of a diverging pair has mutated at a particular site, subsequent mutations in either sequence cannot render the sites any more "different." In fact, subsequent mutations can make them again equal (for example, if a valine mutates to an isoleucine, which mutates back to a valine). Therefore, most distance-based methods correct for such "unseen" substitutions. In practice, application of the rate matrix effectively presumes that some proportion of observed pairwise base identities actually represents multiple mutations and that this proportion increases with increasing overall sequence divergence. Some programs implement, at least optionally, calculation of uncorrected distances, whereas, for example, the MEGA program implements only uncorrected distances for codon and amino acid data. Unless overall divergences are very low, the latter approach is virtually guaranteed to give inaccurate results.

Pairwise distance is calculated using maximum-likelihood estimators of substitution rates. The most popular distance tree-building programs have a limited number of substitution models, but PAUP 4.0 implements a number of models, including the actual model estimated from the data using maximum likelihood, as well as the log-det distance method.

Distance methods are much less computationally intensive than maximum likelihood but can employ the same models of sequence evolution. This is their biggest advantage. The disadvantage is that the actual character data are discarded. The most commonly applied distance-based methods are the unweighted pair group method with arithmetic mean (UPGMA), neighbour joining (NJ), and methods that optimize the additivity of a distance tree, including the minimum evolution (ME) method. Several methods are available in more than one phylogenetics software package but not all implementations allow the same parameter specifications and/or tree optimization features (e.g., branch swapping).

Unweighted pair group method with arithmetic mean (UPGMA)

UPGMA is a clustering or phenetic algorithm—it joins tree branches based on the criterion of greatest similarity among pairs and averages of joined pairs. It is not strictly an evolutionary distance method. UPGMA is expected to generate an accurate topology with true branch lengths only when the divergence is according to a molecular clock or approximately equal to raw sequence dissimilarity. As mentioned earlier, these conditions are rarely met in practice.

Neighbour joining (NJ)

The neighbour-joining algorithm is commonly applied with distance tree building, regardless of the optimization criterion. The fully resolved tree is "decomposed" from a fully unresolved "star" tree by successively inserting branches between a pair of closest (actually,

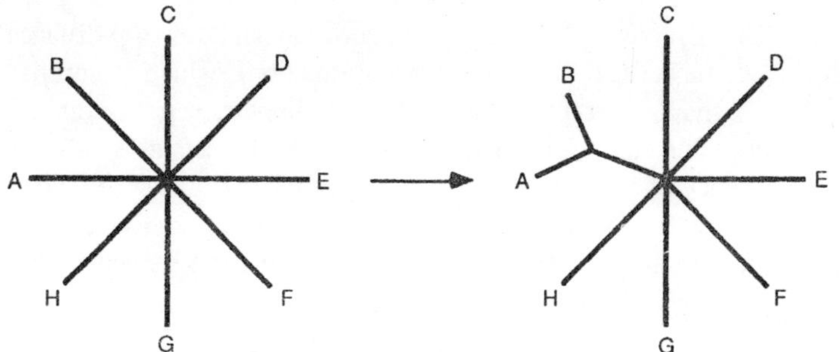

Fig. 9.7. Star decomposition.

most isolated) neighbours and the remaining terminals in the tree. The closest neighbour pair is then consolidated, effectively reforming a star tree, and the process is repeated. The method is comparatively rapid.

Fitch-margoliash (FM)

The Fitch-Margoliash (FM) method seeks to maximize the fit of the observed pairwise distances to a tree by minimizing the squared deviation of all possible observed distances relative to all possible path lengths on the tree. There are several variations that differ in how the error is weighted. The variance estimates are not completely independent because errors in all the internal tree branches are counted at least twice.

Minimum evolution (ME)

Minimum evolution seeks to find the shortest tree that is consistent with the path lengths measured in a manner similar to FM; that is, ME works by minimizing the squared deviation of observed to tree-based distances. Unlike FM, ME does not use all possible pairwise distances and all possible associated tree path lengths. Rather, it fixes the location of internal tree nodes based on the distance to external nodes and then optimizes the internal branch length according to the minimum measured error between these "observed" points. It thus purports to eliminate the nonindependence of FM measurements.

Which distance-based tree-building procedure is best?

ME and FM appear to be the best procedures, and they perform nearly identically in simulation studies. ME is becoming more widely implemented in computer programs, including METREE and PAUP. For protein data, the FM procedure in PHYLIP offers the greatest range of substitution models but no correction for among-site rate heterogeneity. The MEGA and METREE packages include a gamma correction for proteins, but only in conjunction with a raw ("*p*-distance") divergence model (no distance or bias correction), which is unreliable except for small divergences. MEGA also computes separate distances for synonymous and nonsynonymous sites, but this method is valid only in the absence of substitution or base frequency bias and when there is no correction for among-site rate heterogeneity. Thus, for most data sets, using the nucleotide data under a more realistic model might be preferable to MEGA's methods.

Simulation studies indicate that UPGMA performs poorly over a broad range of tree shape space. The use of this method is not recommended; it is mentioned here only because its application seems to persist, as evidenced by UPGMA gene trees appearing in publications.

NJ is clearly the fastest procedure and generally yields a tree close to the ME tree. However, it yields only one tree. Depending on the structure of the data, numerous different trees might be as good or significantly better than the NJ tree.

Character-Based Methods

The character-based methods have little in common with each other, besides the use of the character data at all steps in the analysis. This allows the assessment of the reliability of each base position in an alignment on the basis of all other base positions.

Maximum parsimony (MP)

Maximum parsimony is an optimization criterion that adheres to the principle that the best explanation of the data is the simplest, which in turn is the one requiring the fewest ad hoc assumptions. In practical terms, the MP tree is the shortest—the one with the fewest changes—which, by definition, is also the one with the fewest parallel changes. There are several variants of MP that differ with regard to the permitted directionality of character state change.

To accommodate substitution bias, MP is amenable to weighting; for example, the transformation of a transversion can be weighted relative to a transition. The easiest way to do this is to create a weighting step matrix in which the weights are the reciprocal of the rates estimated using ML as described above. However, step-matrix weighting can greatly slow MP computation.

The MP method performs poorly when there is substantial among-site rate heterogeneity. There are few good fixes for this problem. One approach is to modify the data set to include only sites that exhibit little or no heterogeneity as determined by likelihood estimation. Another approach is to recursively reweight positions according to their propensity to change as observed in preliminary trees. This "successive approximations" approach is automatically facilitated in PAUP, but it is prone to error to the degree that the preliminary trees are incorrect.

MP analyses tend to yield numerous (and sometimes many thousands of) trees that have the same score. Because each is held to be as optimal as any other, only groupings present in the strict consensus of all trees are considered to be supported by the data. The reason that distance and ML tree methods tend to arrive at a single best tree is that their calculations involve division and decimals, whereas MP merely counts discrete steps. For a given data set, a strict consensus of all ME or ML trees that are not significantly worse than optimal probably would yield resolution more or less comparable to the MP consensus. Unfortunately, whereas MP users conventionally present strict consensus (and sometimes consensus of trees one or two steps worse), ME and ML users typically do not.

Simulation studies have shown that MP performs no better than ME and worse than ML when the amount of sequence evolution since lineages diverged is much greater than the amount of divergence that occurred between lineage splits (i.e., in a tree with very long terminal branches and short internal internodes). This condition produces "long branch attraction"—the long branches become artificially connected because the number of nonhomologous similarities the sequences have accumulated exceeds the number of homologous similarities they have retained with their true closest relatives. Character weighting improves the performance of MP under these conditions.

Maximum likelihood (ML)

ML turns the phylogenetic problem inside out. ML searches for the evolutionary model, including the tree itself, that has the highest likelihood of producing the observed data. In practice, ML is derived for each base position in an alignment. The likelihood is calculated in terms of the probability that the pattern of variation at a site would be produced by a particular substitution process, given a particular tree and the overall observed base frequencies. The likelihood becomes the sum of the probabilities of each possible reconstruction of substitutions under a particular substitution process. The likelihoods for all the sites are multiplied to give an overall "likelihood of the tree" (i.e., the probability of the data given the tree and the substitution process). As one can imagine, for one particular tree, the likelihood of the data is low at some sites and high at others. For a "good" tree, many sites will have higher likelihood, so the product of likelihoods is high. For a "poor" tree, the reverse will be true.

The substitution model should be optimized to fit the observed data. For example, if there is a transition bias, evident by an inordinate number of sites that include only purines or pyrimidines, the likelihood of the data under a model that assumes no bias will never be as good as one that does. Likewise, if a substantial proportion of the sites are occupied by a single base and another substantial proportion have equal base frequencies, the likelihood of the data under a model that assumes that all sites evolve equally will be less than that of a model that allows rate heterogeneity. Modifying the substitution parameters, however, modifies the likelihood of the data associated with particular trees. Thus, the tree yielding the highest likelihood under one substitution model might yield much lower likelihood under another.

Because ML uses great amounts of computational time, it is usually impractical to perform a complete search that simultaneously optimizes the substitution model and the tree for a given data set. An economical, heuristic approach is recommended. Perhaps the best time saver in this regard is preliminary ML estimation of the substitution model (as can be performed using PAUP). This procedure can be applied iteratively, searching for better ML trees, then reestimating the parameters, and then searching for better trees.

As algorithms, computers, and phylogenetic understanding have improved, the ML criterion has become more popular for molecular phylogenetic analysis. In simulation studies, ML has consistently outperformed ME, and MP when the data analysis proceeds according to the same model that generates the data. ML will always be the most computationally intensive method of all, however, so there will always be situations in which it is not practical.

Searching for Trees

The number of unique phylogenetic trees increases exponentially with the number of taxa, becoming astronomical even for, say, 50 sequences. In most cases, computational limitations permit exploration of only a small fraction of possible trees. The exact number will depend mainly on the number of taxa, the optimality criterion (e.g., MP is much faster than ML), the parameters (e.g., unweighted MP is much faster than weighted; ML with fewer preset parameters is much faster than with more and/or simultaneously optimized parameters), computer hardware, and computer software (some algorithms are faster than others; some software allows multiprocessing; some software limits the number and kind of trees that can be stored in memory). The search procedure is also affected by data structure: poorly resolvable data produce more "nearly optimal" trees that must be evaluated to find the most optimal.

Branch-swapping algorithms successively modify existing trees built by an initial step. The algorithms range from those that generate all possible unique trees (exhaustive algorithms) to those that evaluate only minor modifications.

Quartet puzzling is a relatively rapid tree-searching algorithm available for ML tree building and is available in PUZZLE.

One of the best ways to economize the search effort is to prune the data set. For example, it might be apparent from the data alone or from preliminary searching that a particular cluster of five terminals is unresolvable, that the arrangement of these terminals does not impact the remainder of the topology, and/or that resolution of these terminals is not the objective of the analysis. Removing four of the terminals from the analysis simplifies the search by several orders of magnitude.

Every analysis is unique. The elements that influence the choice of optimal search strategy (amount of data, structure of data, amount of time, hardware, objective of analysis) are too variable to suggest a foolproof recipe. Thus, researchers must be familiar with their data; they must also have specific objectives in mind, understanding the various search procedures as well as the capabilities of their hardware and software.

Rooting Trees

The methods described above produce unrooted trees (i.e., trees having no evolutionary polarity). To evaluate evolutionary hypotheses, it is often necessary to locate the root of the tree. Rooting phylogenetic trees is not a trivial problem.

If one accepts a molecular clock, then the root will always be at the midpoint of the longest span across the tree. Whether molecular evolution is indeed clocklike generally remains a contentious issue, but most gene trees exhibit unclocklike behaviour regardless of where the root is placed. Thus, rooting is generally evaluated by extrinsic evidence, that is, by means of determining where the tree would attach to an "outgroup," which can be any organism/sequence not descended from the nearest common ancestor of the organisms/sequences analyzed (for example, a bird sequence could be used to root an analysis of mammals). Outgroup rooting, however, creates a dilemma: an outgroup that is closely related to the ingroup might be simply an erroneously excluded member of the ingroup. A clearly distant outgroup (e.g., a fungus for an analysis of plants) can have a sequence so diverged that its attachment to the ingroup is subject to the long-branch attraction problem mentioned above. It is wise to examine the results obtained for trees both with and without an outgroup.

Another means of rooting involves analysis of a duplicated gene or gene with an internal duplication. If all the paralogs from most or all of the organisms are included in the analysis, then one can logically root the tree exactly where the paralog gene trees converge, assuming that there are not long branch problems in all trees.

Tree Evolution

Several procedures are available that evaluate the phylogenetic signal in the data and the robustness of trees. The most popular of the former class are tests of data signal versus randomized data (skewness and permutation tests). The latter class includes tests of tree support from resampling of observed data (nonparametric bootstrap). The likelihood ratio test provides a means of evaluating both the substitution model and the tree.

Randomized Trees (Skewness Test)

Simulation studies indicate that the distribution of random MP tree lengths generated using random data sets will be symmetrical, whereas those using data sets with phylogenetic signal will be skewed. The critical value of the g_1 statistic of skewness will vary with the number of taxa and variable sites in the sequence. The test does not estimate the reliability of a particular topology, and it is sensitive to even very small amounts of signal present in an otherwise random data set. If taxa from groups that are obviously well supported by the data are selectively deleted, the test can be used to determine whether a phylogenetic signal remains, provided at least 10 variable characters and 5 taxa are examined. The procedure is implemented in PAUP.

Randomized Character Data (Permutation Tests)

The randomized data approach determines whether an MP tree or portion of it derived from the actual data could have arisen by chance. The data are not truly randomized but permuted within each aligned column, so that covariation in the initial data is removed. The result is an alignment of sequences that are not random sequences; rather, the base at each site in these sequences is randomly drawn from the population of bases occupying that site in the overall alignment. The permutation tail probability test (PTP) compares the score for the MP tree with trees generated by numerous permutations of the data at each site, determining only whether there is a phylogenetic signal in the original data. A topology-dependent test (T-PTP) compares the scores for specific trees to determine whether the difference can be attributed to chance. This method does not evaluate whether the tree or any portion of it is correct. In particular, the T-PTP test will appear to corroborate groups that are in trees close to the MP tree but not in it. This is because the method detects the collective signal that places a taxon even approximately, if not actually, in its correct position. The results can be fine-tuned, however, by additional applications using relevant subsets of the data. The procedure is implemented in PAUP.

Bootstrap

Bootstrapping is a resampling tree evaluation method that works with distance, parsimony, likelihood, and just about any other tree derivation method. It was invented in 1979 and introduced as a tree evaluation method in phylogenetic analysis by Felsenstein (1985). The result of bootstrap analysis is typically a number associated with a particular branch in the phylogenetic tree that gives the proportion of bootstrap replicates that supports the monophyly of the clade.

How is this done practically? Bootstrapping can be considered a two-step process comprising the generation of (many) new data sets from the original set and the computation of a number that gives the proportion of times that a particular branch (e.g., a taxon) appeared in the tree. That number is commonly referred to as the bootstrap value. New data sets are created from the original data set by sampling columns of characters at random from the original data set with replacement. "With replacement" means that each site can be sampled again with the same probability as any of the other sites. As a consequence, each of the newly created data sets has the same number of total positions as the original data set, but some positions are duplicated or triplicated and others are missing. It is therefore possible that some of the newly created data sets are completely identical to the original set—or, on the other extreme, that only one of the sites is replicated, say, 500 times, whereas the remaining 499 positions in the original data set are dropped.

Although it has become common practice to include bootstrapping as part of a thorough phylogenetic analysis, there is some discussion on what exactly is measured by this method. It was originally suggested that the bootstrap value is a measure of repeatability. In more recent interpretations, it has been considered to be a measure of accuracy—a biologically more relevant parameter that gives the probability that the true phylogeny has been recovered. On the basis of simulation studies, it has been suggested that, under favourable conditions (roughly equal rates of change, symmetric branches), bootstrap values greater than 70% correspond to a probability of greater than 95% that the true phylogeny has been found. By the same token, under less favourable conditions, bootstrap values greater than 50% will be overestimates of accuracy. Simply put, under certain conditions, high bootstrap values can make the wrong phylogeny look good; therefore, the conditions of the analysis must be considered. Bootstrapping can be used in experiments in which trees are recomputed after internal branches are deleted one at a time. The results provide information on branching orders that are ambiguous in the full data set.

Parametric Bootstrap

The parametric bootstrap differs from the nonparametric in that it uses simulated (yet actual) replicates rather than pseudoreplicates. In the case of phylogenetic sequence analysis, replicate data sets of the same size as the original data set are generated according to a specified model of sequence evolution, including the optimal tree topology determined according to that model. Each data set is then analyzed according to the method of interest. Support for the branches in the test tree can be determined in much the same way as in the nonparametric bootstrap.

Likelihood Ratio Tests

As the name implies, likelihood ratio tests are applicable to ML, analyses. A suboptimal likelihood value is evaluated for significance against a normal distribution of the error in the optimal model. In ideal applications, the error curve is presumed to be a X^2 distribution. Thus, the test statistic is twice the difference between the optimal and test values, and the degrees of freedom is the number of parameter differences. Application of the X^2 test to alternative phylogenetic trees is problematic, especially because of the "irregularity of [the] parameter space", but its use has been advocated for evaluating optimality of the substitution model when the number of parameters between models is known.

Phylogenetics Software

PHYLIP and PAUP compete as the most widely used phylogenetic analysis software, although other newer applications such as PUZZLE are beginning to compete. Here, PHYLIP and PAUP will be described in the most detail, with references made to other available packages that have useful features. However, the number of programs available is now so numerous, many each having their own useful features, that the reader is referred to the list of Internet resources at the end of this chapter for further information.

PHYLIP

PHYLIP (for phylogeny inference package) is a package now consisting of about 30 programs that cover most aspects of phylogenetic analysis. PHYLIP is free and available for a wide variety of computer platforms (Mac, DOS, UNIX, VAX/VMS, and others). According to its author, PHYLIP is currently the most widely used phylogeny program.

PHYLIP is a command-line program and does not have a point-and-click interface, as programs like PAUP do. The documentation is well written and very comprehensive, and the interface is straightforward. A program within PHYLIP is invoked by typing its name, which automatically causes the data to be read from a file called "infile" or a file name you specify if no infile exists. This infile must be in PHYLIP format; this format is clearly described in the documentation, and most sequence analysis programs offer the ability to export sequences in this format. For example, if an alignment is produced using CLUSTAL W or edited using GeneDoc, the alignment may be saved in PHYLIP format and then used in PHYLIP programs directly. Once the user activates a given PHYLIP program and loads the infile, the user can then choose from an option menu or accept the default values. The program will write its output to a file called "outfile" (and "treefile" where applicable). If the output is to be read by another program, "outfile" or "treefile" must be renamed before execution of the next program, as all files named outfile/tree file in the current directory are overwritten at the beginning of any program execution. The tree file generated is a widely used format that can be imported into a variety of tree-drawing programs, including DRAWGRAM and DRAWTREE that come with this package. However, these PHYLIP tree-drawing programs produce low-resolution graphics, so a program such as TreeView is instead recommended. Particulars of some of the PHYLIP tree-inference programs are discussed below.

PROTDIST is a program that computes a distance matrix for an alignment of protein sequences. It allows the user to choose between one of three evolutionary models of amino acid replacements. The simplest, fastest (and least realistic) model assumes that each amino acid has an equal chance of turning into 1 of the other 19 amino acids. The second is a category model in which the amino acids are redistributed among different groups; transitions in this model are evaluated differently depending on whether the change would result in an amino acid in the same or in a different group. The third (default) method, which is recommended, uses a table of empirically observed transitions between amino acids, the Dayhoff PAM 001 matrix.

NEIGHBOUR is a tree-generating program that utilizes the distance matrix data generated from a program such as PROTDIST and generates a tree using the neighbour-joining method. This is one of the more popular methods, due to its speed of computation.

FITCH is another tree-generating program similar to NEIGHBOUR but much more robust. It also uses distance matrix data, such as that described in PROTDIST, and generates a tree using the method of Fitch-Margoliash. This method, while more robust than NEIGHBOUR, tends to produce a similar final answer, yet takes longer to compute. Although computational times are often significantly longer, the quality of the results produced by the method often makes this method the method of choice in these types of analyses.

PROTPARS is a parsimony program for protein sequences that generates trees without utilizing a distance matrix. The evolutionary model is different from the ones used in the PROTDIST program in that it considers the underlying changes in the nucleotide sequence to evaluate the probabilities of the observed amino acid changes. Specifically, it makes the (biologically meaningful) assumption that synonymous changes [e.g., GCA (alanine) → GCC (alanine)] occur more often than nonsynonymous changes. As a consequence, a transition between two amino acids that would require, for example, three nonsynonymous changes in the underlying nucleotide sequences, is assigned a lower probability than an amino acid change calling for two nonsynonymous changes and one synonymous change. PROTPARS does not have an option that uses empirical values for amino acid changes (e.g.. PAM matrices).

DNADIST computes a distance matrix from nucleotide sequences. Trees are generated by running the output through NEIGHBOUR or other distance matrix programs in the PHYLIP package. DNADIST allows the user to choose between three models of nucleotide substitution. The older Jukes and Cantor model is similar to the simple model in the PROTDIST program in that it assumes equal probabilities for all changes. The more recent Kimura two-parameter model is very similar but allows the user to weigh transversion more heavily than transitions. PHYLIP also comprises DNAML, a maximum-likelihood program for nucleotide data. Because the program is fairly slow, the use of its faster "sibling," the fastDNAml program described below, is recommended.

SEQBOOT and CONSENSE are required for bootstrap analysis. SEQBOOT is used to generate any number of replicates of the data;

these replicates are then used in programs within the PHYLIP suite for analysis. The resulting tree file contains as many trees as there are replicates of the data, so this file needs to be run through CONSENSE to generate the consensus tree from the analysis.

PAUP

The objective of the development of PAUP is to provide a phylogenetics program that includes as many functions (including tree graphics) as possible in a single, platform-independent program with a menu interface. PAUP stands for phylogenetic analysis using parsimony and still contains one of the most sophisticated parsimony programs available. Version 3 performed only MP-associated tree-building and analytical functions. PAUP version 4 also includes distance and ML functions for nucleotide data and other new features.

Current tree-building functions in PAUP include MP, and, for nucleotide data, distance and ML using the fastDNAml algorithm. In addition, PAUP performs Lake's method of invariants. Each tree-building program permits a variety of options. The MP options include specification of any character-weighting scheme. Distance options include choice of NJ, ME, FM, and UPGMA procedures. The full range of options and their current values can be examined using the menu and/ or by typing pse[ttings]?, dse[ttings]?, and lse[ttings] ? for parsimony, distance, and likelihood, respectively. Both distance and ML allow detailed specification of the substitution model (values of substitution, gamma, and invariant-sites parameters, assuming equal, specified, or empirical base frequencies), and these can be estimated for any tree by setting the parameter values to est[imate] and applying the des[cribe tree] command with a desired tree in memory.

According to the release notes accompanying PAUP test version 4, PAUP* *usually* finds trees with likelihoods as high or higher [i.e., better] than PHYLIP (both because PAUP*'s tree rearrangements are more extensive and because its convergence criterion for branch-length iteration is stricter).

With any tree-building method, PAUP allows a variety of tree search options. These include algorithm specification for generating the initial tree (starting tree): NJ, stepwise addition, or input tree(s). The stepwise-addition algorithm allows numerous options, including addition of taxa "as is" (taxa added in file order): closest, furthest, or random with any number of replicates. All the stepwise options allow for any maximal number of partial trees to be retained and built on during taxon addition. Increasing this number to, say, 100, is another means of increasing the diversity of starting topologies, although these are not random.

A random addition strategy provides a useful complement to the default search strategy (closest addition, TBR swapping, saving all best trees). In the random search, a large number of replicates can be combined with the faster NNI swapping algorithm. For MP analysis, in which a large number of equal-length trees might exist, the search should specify saving from each replicate only a few trees that match or are better than the score of the slower search. In addition, the number of suboptimal trees (the trees that will be swapped on to find better trees) should be limited by setting MAXTREES to a low number (e.g., 10). By using this strategy to explore areas of "tree space" possibly missed in the slow search, one sometimes finds better trees and/or additional unique optimal trees.

PAUP performs the nonparametric bootstrap for distance, MP, and ML, using all options available for tree building with these methods. When a bootstrap or jackknife with MP is under way, MAXTREES should be set between 10 and no more than 100. This is because poorly resolvable portions of an MP tree will usually be even less resolvable with resampled data; hence, a replicate could find astronomical numbers of equal-length trees. Because tree branches weakly supported by the full data set will not have high bootstrap or jackknife values, limiting MAXTREES will have little, if any, bearing on the results, especially if the number of replicates is increased to, say, 1,000.

In addition, PAUP performs the Kishino-Hasegawa test to compare MP or ML trees, computes four types of consensus of multiple trees (usually used for multiple equal-length MP trees), computes stepwise differences between MP trees, and evaluates signal conflicts between specified partitions of sites (e.g., nuclear and organellar sequence data in a combined analysis).

Other Programs

In addition to PAUP and PHYLIP, there are phylogenetics programs that have some unique capabilities but are generally more limited in their procedures and portability. These include FastDNAml, PUZZLE, MACCLADE, and MOLPHY.

FastDNAml

FastDNAml is a freestanding maximum-likelihood, tree-building program. Although it is currently not part of the PHYLIP package, it uses largely the same input and output conventions, and the results of fastDNAml and PHYLIP's DNAML should be very similar or identical. FastDNAml can be run on parallel processors, and it comes with a number of useful scripts (in particular for bootstrapping and jumbling the sequence input order). To take full advantage of the program, knowledge of UNIX is beneficial. The source code for UNIX systems is publicly available from the RDP Web site, and a Power Macintosh version is available by FTP.

PUZZLE or TREE-PUZZLE

PUZZLE or TREE-PUZZLE, as it is now called, is a maximum likelihood-based program that implements a fast tree search algorithm (quartet puzzling) that allows analysis of large data sets and automatically assigns estimations of support to each internal branch. PUZZLE also computes pairwise maximum-likelihood distances as well as branch lengths for user-specified trees. PUZZLE also offers a novel method, likelihood mapping, to investigate the support of a hypothesized internal branch without computing an overall tree and to visualize the phylogenetic content of a sequence alignment. It conducts a number of statistical tests (X^2 test for homogeneity of base composition, likelihood ratio clock test, Kishino-Hasegawa test) and includes a large range of models of substitution. Rate heterogeneity is modeled by a discrete gamma distribution and by allowing invariable sites.

MACCLADE

MACCLADE is an interactive Macintosh program for manipulating trees and data and studying the phylogenetic behaviour of characters. It uses the NEXUS file format and will read PAUP data and tree files. Some information in PAUP files will be ignored in MACCLADE (e.g., gap mode), but information in a PAUP "assumptions" block will be imported, including character weightings and character and taxon sets. Several subtle differences exist between PAUP and MACCLADE files. Thus, PAUP files edited with MACCLADE and vice versa should be saved under new names and the unedited file maintained separately. PHYLIP, NBRF-PIR, and text files are also readable by MACCLADE. Any method can be used to generate the trees, but MACCLADE's functions are based strictly on parsimony. For example, the program allows one to trace the evolution of each individual character on any tree. The MP and ML reconstruction functions differ, however, and the ML function is considered more realistic. Tree topologies can be manipulated by dragging branches, and flipping branches can produce aesthetic modifications in tree symmetry.

MACCLADE includes additional features relevant to sequence analysis, including a chart of character number versus number of changes in a tree, which is useful for visualizing among-site rate heterogeneity, and a chart of the overall numbers of changes from one base to another over an MP tree ("state changes and stasis" chart: the values are sometimes erroneously reported in the literature as substitution "rates," but there is no correction for branch lengths or among-site rate heterogeneity).

MOLPHY

MOLPHY is a shareware package of programs and utilities for ML analysis and statistics of nucleotide or amino acid sequences. It has been tested on Sun OS and HP9000/700 systems. Practical application requires some knowledge of UNIX file management. The ML procedures are similar to those in PHYLIP, but there is a wider range of amino acid substitution models and options for faster, heuristic searches, including an option to use "local bootstrap" analyses (i.e., a bootstrap on subtrees under the assumption that the remainder of the tree is correct) to search for better ML trees. The output includes branch-length estimates and standard error. Analysis of separate codon positions is possible. MOLPHY uses a subset of the nucleotide substitution models available in PAUP, although it allows user-specified parameter values. The current MOLPHY lacks a bootstrap option and also has no accommodation for among-site rate heterogeneity.

Tree drawing

There are a number of tree-drawing programs available now, such as TreeTool (X-windows), TreeDraw (Macintosh), PHYLODENDRON (Macintosh), TreeView (Macintosh, Microsoft Windows), or the tree-drawing tool in PAUP, and all handle standard tree files. These programs facilitate not only the generation of trees suitable for publication or other presentation but also facilitate viewing of the data in general. For example, programs such as the freely available TreeView enable the user to manipulate the view of branching order, root the tree, and other graphical manipulations that aid the user.

Internet-Accessible Phylogenetic Analysis Software

Currently, there are few Web-based applications that will permit an investigator to perform phylogenetic analyses over the Web. However, these kinds of resources are appearing in increasing numbers, and, presumably as the Internet bandwidth increases and servers have faster CPUs, this may become even more common. Highlighted here are three Internet-based applications that provide phylogenetic analysis capabilities: WEBPHYLIP, PhyloBLAST, and the "BLAST2 & Orthologue Search Server." These illustrate the variety of applications currently available. Although all use PHYLIP programs, the latter two combine phylogenetic analysis with BLAST to aid the user in retrieving sequences for analysis.

WEBPHYLIP

WEBPHYLIP uses CGI/Perl scripts to produce a Web-based cut-and-paste interface to the PHYLIP programs. Unfortunately, the programs are not linked together; therefore, to generate a neighbour-joining tree, for example, the user must run multiple analyses (PROTDIST and NEIGHBOUR in this case). Analyses may time-out if too intensive of an analysis is requested. Also, the trees cannot be easily viewed; for example, if the user has a PC, they must have ghostview or another such PostScript viewer installed to actually view the results. However, the Web site provides excellent flowcharts and other helpful documentation about running the programs, and an extensive collection of the PHYLIP programs is available.

PhyloBLAST

PhyloBLAST is also based on CGI/Perl scripts. PhyloBLAST compares a user's protein sequence to the SWISS-PROT/TREMBL databases using BLAST2 and then allows user-defined phylogenetic analyses to be performed on selected sequences from the BLAST output. Neighbour-joining and parsimony analyses may be performed, either with or without bootstrapping, using PHYLIP programs. Flexible features, such as the ability to input premade multiple sequence alignments and use all options found in the PHYLIP programs, provide additional functionality that goes beyond the simple analysis of a BLAST result. Because PHYLIP programs need to generate trees that are linked, there is less input required by the user than for WEBPHYLIP; however, DNA analysis and analysis using some programs (for example, FITCH) are not currently available. However, PhyloBLAST's ability to generate trees-containing hyperlinks to further protein sequence information or generate JPEG graphics of trees is a considerable advantage. Also, the program is set up to handle Web page time-outs so long analyses are not a problem (and can be E-mailed to the user if preferred).

BLAST2 and Orthologue Search Server

This is a fairly specialized application of phylogenetic analysis of a BLAST output for the identification of orthologs verses paralogs. It is based on the use of CLUSTAL W, WU-BLAST2, and the tree-reconciling algorithm of Page (1994). This tool first performs a BLAST analysis and then performs a phylogenetic analysis on user-selected sequences based on a CLUSTAL W alignment and PHYLIP's neighbour-joining methods. This resulting neighbour-joining tree ("gene tree") is ultimately compared with a predicted species tree and the reconciled tree viewed for analysis. The philosophy here is that, whenever the phylogeny of the species matches the phylogeny of the gene tree, these genes will be deemed orthologous.

Although this is a useful tool, users should be cautioned that this does not represent a comprehensive phylogenetic analysis, due to the automated nature of the application. Its use should be primarily as intended: to gain insight into what homologous sequences are orthologous in an automated fashion. Further analysis should be performed for any particularly in-depth investigation using less automated alignment and phylogenetic analysis that suits the sequences being investigated.

Some Simple Practical Considerations

1. Paradoxical as it may sound, by far the most important factor in inferring phylogenies is not the method of phylogenetic inference but the quality of the input data. The importance of data selection and in particular of the alignment process cannot be overestimated. Even the most sophisticated phylogenetic inference methods are not able to correct for erroneous input data.
2. Look at the data from as many angles as possible. Use each of the three main methods (distance, maximum parsimony, maximum likelihood) and compare the resulting trees for consistency. At the same time, be aware that one cannot rely on having arrived at a good estimate for the true phylogeny just because all three methods produce the same tree. Unfortunately, consistency among results obtained by different methods does not necessarily mean that the result is statistically significant (or represents the true phylogeny), since there can be several reasons for such correspondence.
3. The choice of outgroup taxa can have as much influence on the analysis as the choice of ingroup taxa. Complication will occur in particular when the outgroup shares an unusual property (e.g., composition bias or clock rate) with one or several ingroup taxa. It is therefore advisable to compute every analysis with several outgroups and check for congruency of the ingroup topologies.
4. Be aware that programs can give different answers (trees) depending on the order in which the sequences appear in the input file.

Table 9.1. Important internet resources.

Compilation of available phylogeny program	*http://evolution.genetics. washington.edu/phylip/software.html*
BLAST2 & Orthologue Search	*http://www.Bork.EMBL-Heidelberg.DE/Blast2e/*
CLUSTAL W	*http://www-igbmc.u-strasbg.fr/BioInfo/*
GeneDoc	*http://www.psc.edu/biomed/genedoc/*
GeneTree	*http://taxonomy.zoology.gla.ac.uk/rod/genetree/genetree.html*
PHYLIP	*http://evolution.genetics.washington.edu/phylip.html*
PhyloBLAST	*http://www.pathogenomics.bc.ca/phyloBLAST/*
Phylogenetic Resources	*http://www.ucmp.berkeley.edu/subway/phylogen.html*
PUZZLEBOOT	*http://www.tree-puzzle.de*
ReadSeq	*http://dot.imgen.bcm.tmc.edu:9331/seq-util/Options/readseq.html*
RDP Tree	*http://rdp.life.uiuc.edu/RDP/commands/sgtree.html*
TreeView	*http://taxonomy.zoology.gla.ac.uk/rod/treeview.html*
WebPHYLIP	*http://sdmc.krdl.org.sg.8080/~lxzhang/phylip/*
WHS	*http://www.cladistics.org/education.html*

PHYLIP, PAUP and other phylogenetic software provide a "jumble" option that reruns the analysis with different (jumbled) input orders. If for whatever reason the tree must be computed in a single run, sequences that are suspected of being "problematic" should be placed toward the end of the input file, lo lower the probability that tree rearrangement methods will be negatively influenced by a poor initial topology stemming from any problematic sequences.

10

DATABASE SEARCHING

As discussed earlier, GenBank was created in response to the explosion in sequence information resulting from a panoply of scientific efforts such as the Human Genome Project. To review, GenBank is an annotated collection of all publicly available DNA and protein sequences and is maintained by the National Center for Biotechnology Information (NCBI). As of this writing, GenBank contains 7 million sequence records covering almost 9 billion nucleotide bases. Sequences find their way into GenBank in several ways, most often by direct submission by individual investigators through tools such as Sequin or through "direct deposit" by large genome sequencing centers.

GenBank, or any other biological database for that matter, serves little purpose unless the database can be easily searched and entries retrieved in a usable, meaningful format. Otherwise, sequencing efforts have no useful end, since the biological community as a whole cannot make use of the information hidden within these millions of bases and amino acids. Much effort has gone into making such data accessible to the average user, and the programs and interfaces resulting from these efforts are the focus of this chapter. The discussion centers on querying the NCBI databases because these more "general" repositories are far and away the ones most often accessed by biologists, but attention is also given to a number of smaller, specialized databases that provide information not necessarily found in GenBank.

INTEGRATED INFORMATION RETRIEVAL: THE ENTREZ SYSTEM

The most widely used interface for the retrieval of information from biological databases is the NCBI Entrez system. Entrez capitalizes on the fact that there are preexisting, logical relationships between the individual entries found in numerous public databases. For example, a paper in MEDLINE (or, more properly, PubMed) may describe the sequencing of a gene whose sequence appears in GenBank. The nucleotide sequence, in turn, may code for a protein product whose sequence is stored in the protein databases. The three-dimensional structure of that protein may be known, and the coordinates for that structure may appear in the structure database. Finally, the gene may have been mapped to a specific region of a given chromosome, with that information being stored in a mapping database. The existence of such natural connections, mostly biological in nature, argued for the development of a method through which all information about a particular biological entity could be found without having to sequentially visit and query disparate databases. Entrez, to be clear, is not a database itself—it is the interface through which all of its component databases can be accessed and traversed. The Entrez information space includes PubMed records, nucleotide and protein sequence data, threedimensional structure information, and mapping information. The strength of Entrez lies in the fact that *all* of this information can be accessed by issuing one and only one query. Entrez is able to offer integrated information retrieval through the use of two types of connection between database entries: *neighbouring* and *hard links*.

Neighbouring

The concept of neighbouring allows for entries *within* a given database to be connected to one another. If a user is looking at a particular PubMed entry, the user can ask Entrez to find all other papers in PubMed that are similar in subject matter to the original paper. Similary, if a user is looking at a sequence entry, Entrez can return a list of all other sequences that bear similarity to the original sequence. The establishment of neighbouring relationships within a database is based on statistical measures of similarity, as follows.

BLAST

Sequence data are compared with one another using the Basic Local Alignment Search Tool or BLAST. This algorithm attempts to find "high-scoring segment pairs" (HSPs), which are pairs of sequences that can be aligned with one another and, when aligned, meet certain scoring and statistical criteria.

VAST

Sets of coordinate data are compared using a vector-based method known as VAST, for Vector Alignment Search Tool. There are three major steps that take place in the course of a VAST comparison:

1. First, based on known three-dimensional coordinate data, all of the α-helices and β-sheets that comprise the core of the protein are identified. Straight-line vectors are then calculated based on the position of these secondary structure elements. VAST keeps track of how one vector is connected to the next (that is, how the C-terminal end of one vector connects to the N-terminal end of the next vector), as well as whether a particular vector represents an α-helix or a β-sheet. Subsequent steps use *only* these vectors in making comparisons to other proteins. In effect, most of the coordinate data is discarded at this step. The reason for this apparent oversimplification is simply due to the scale of the problem at hand; with over 11,000 structures in PDB, the time that it would take to do an in-depth comparison of each and every structure with all of the other structures in the database would make the calculations both impractical and intractable. The user should keep this simplification in mind when making biological inferences based on the results presented in a VAST table.
2. Next, the algorithm attempts to optimally align these sets of vectors, looking for pairs of structural elements that are of the same type and relative orientation, with consistent connectivity between the individual elements. The object is to identify highly similar "core substructures," pairs that represent a statistically significant match above that which would be obtained by comparing randomly chosen proteins with one another.
3. Finally, a refinement is done using Monte Carlo methods at each residue position in an attempt to optimize the structural alignment.

Through this method, it is possible to find structural (and, presumably, functional) relationships between proteins in cases that may lack overt sequence similarity. The resultant alignment need not be global; matches may be between individual domains of different proteins.

It is important to note here that VAST is not the best method for determining structural similarities. More robust methods, such as homology model building, provide much greater resolving power in determining such relationships, since the raw information within the three-dimensional coordinate file is used to perform more advanced calculations regarding the positions of side chains and the thermodynamic nature of the interactions between side chains. Reducing a structure to a series of vectors *necessarily* results in a loss of information. However, considering the magnitude of the problem here—

again, the number of pairwise comparisons that need to be made—and both the computing power and time needed to employ any of the more advanced methods, VAST provides a simple and fast first answer to the question of structural similarity.

Weighted key terms

The problem of comparing sequence data somewhat pales next to that of comparing PubMed entries, free text whose rules of syntax are not necessarily fixed. Given that no two people's writing styles are exactly the same, finding a way to compare seemingly disparate blocks of text poses a substantial problem. Entrez employs a method known as the relevance pairs model of retrieval to make such comparisons, relying on what are known as weighted key terms. This concept is best described by example. Consider two manuscripts with the following titles:

```
BRCA1 as a Genetic Marker for Breast Cancer
Genetic Factors in the Familial Transmission of the
Breast Cancer BRCA1 Gene
```

Both titles contain the terms BRCA1, Breast, and Cancer, and the presence of these common terms may indicate that the manuscripts are similar in their subject matter. The proximity between the words is also taken into account, so that words common to two records that are closer together are scored higher than common words that are further apart. In the current example, the terms Breast and Cancer would score higher based on proximity than either of those words would against BRCA1, since the words are next to each other. Common words found in a title are scored higher than those found in an abstract, since title words are presumed to be "more important" than those found in the body of an abstract. Overall weighting depends on the frequency of a given word among all the entries in PubMed, with words that occur infrequently in the database as a whole carrying a higher weight. Regardless of the method by which the neighbouring relationships are established, the ability to actually code and maintain these relationships is rooted in the format underlying all of the constituent databases. This format, called Abstract Syntax Notation (ASN.1), provides a format in which all similar fields are all structured identically, regardless of whether the entry is in a protein database, nucleotide database, and so forth.

Hard Links

The hard link concept is much easier conceptually than neighbouring. Hard links are applied between entries in different databases and exist

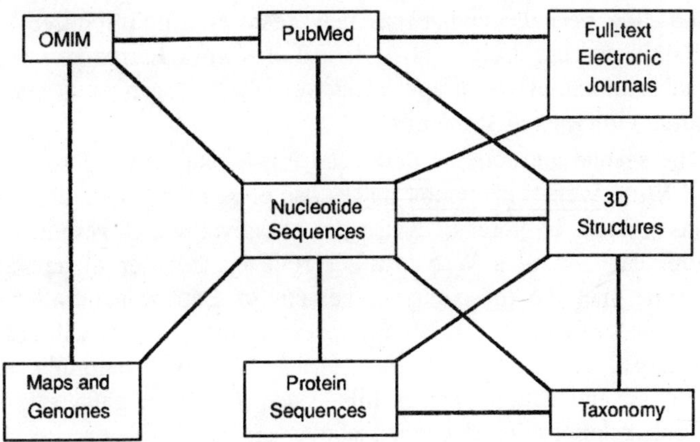

Fig. 10.1. Overview of the relationships in the Entrez integrated information retrieval system.

everywhere there is a logical connection between entries. For instance, if a PubMed entry talks about the sequencing of a cosmid, a hard link is established between the PubMed entry and the corresponding nucleotide entry. If an open reading frame in that cosmid codes for a known protein, a hard link is established between the nucleotide entry and the protein entry. If, by sheer luck, the protein entry has an experimentally deduced structure, a hard link would be placed between the protein entry and the structural entry.

As suggested by the figure, searches can, in essence, begin anywhere within Entrez—the user has no constraints with respect to where the foray into this information space must begin. However, depending on which database is used as the jumping-off point, different fields are available for searching. This stands to reason, inasmuch as the entries in databases of different types are necessarily organized differently, reflecting the biological nature of the entity they are trying to catalog.

Implementations

Regardless of platform, Entrez searches can be performed using one of two interfaces. The first is a client-server implementation known as Network Entrez. This is the fastest of the Entrez programs in that it makes a direct connection to an NCBI "dispatcher." The graphical user interface features a series of windows, and each time a new piece of information is requested, a new window appears on the user's screen. Because the client software resides on the user's machine, it is up to the user to obtain, install, and maintain the software, downloading periodic updates as new features are introduced. The installation process itself is fairly trivial. Network Entrez also comes bundled with interactive, graphical viewers for both genome sequences and three-dimensional structures.

The second and more widely used implementation is through the World Wide Web. This option makes use of available Web browsers, such as Internet Explorer or Netscape, to deliver search results to the desktop. The use of a Web browser relieves the user of having to make sure that the most current version of Entrez is installed—as long as the browser is of relatively recent vintage, results will always be displayed via the latest Entrez release. The Web naturally lends itself to an application such as this, since all the neighbouring and hard link relationships described above can easily be expressed as hypertext, allowing the user to navigate by clicking on selected words in an entry.

The advantage of the Web implementation over the Network version is that the Web allows for the ability to link to external data sources, such as full-text versions of papers maintained by a given journal or press or specialized databases that are not part of Entrez proper. The speed advantage that is gained by the network version causes its limitation in this respect; the direct connection to the NCBI dispatcher means that the user, once connected to NCBI, cannot travel anywhere else. The other main difference between the two methods lies simply in the presentation: the Network version uses a series of windows in presenting search results, whereas the Web version is formatted as sequential pages, following the standard Web paradigm. The final decision is one of personal preference, for both methods will produce the same results *within* the Entrez search space. However, given that Web Entrez can link to external data sources, the remainder of this discussion will focus on the Web implementation.

Entrez Discovery Pathway: Examples

The best way to illustrate the integrated nature of the Entrez system and to drive home the power of neighbouring is by considering two biological examples, using the Web version of Entrez as the interface.

The simplest way to query Entrez is through the use of individual search terms, coupled together by Boolean operators such as AND, OR, or NOT. Consider the case in which one wants to retrieve all papers that discuss aspirin in the context of treating or preventing atherosclerosis. Beginning at the Entrez home page, one would select

PubMed from the Search pull-down menu to indicate that the search is to take place in the bibliographic portion of the Entrez search space. Within the text box to the right, one would simply type atherosclerosis [MH] AND aspirin [NM]. The [MH] qualifying the first term indicates to Entrez that this is a *MeSH term*; MeSH stands for *medical subject heading* and is the qualifier that should be used when searching by subject. The [NM] qualifying the second term indicates that this is the name of a substance or chemical. The user can further narrow down the query by adding additional terms, if the user is interested in a more specific aspect of the pharmacology or if there are quite simply too many papers to read.

At this point, to actually look at one of the papers resulting from the search, the user can click on a hyperlinked author's name. By doing so, the user is taken to the Abstract view for the selected paper. The Abstract view presents the name of the paper, the list of authors, their institutional affiliation, and the abstract itself, in standard format. A number of alternative formats are available for displaying this information, and these various formats can be selected using the pull-down menu next to the Display button. Switching to Citation format would produce a very similar-looking entry, the difference being that cataloguing information such as MeSH terms and indexed substances relating to the entry are shown below the abstract. MEDLINE format produces the MEDLINE/MEDLARS layout, with two-letter codes corresponding to the contents of each field going down the left-hand side of the entry (e.g., the author field is denoted by the code AU). Entries in this format can be saved and easily imported into third-party bibliography management programs, such as EndNote and Reference Manager.

At the top of the entry are a number of links that are worth mentioning. First, on the right-hand side is a hyperlink called Related Articles. This is one of the entry points from which the user can take advantage of the neighbouring and hard link relationships described earlier. If the user clicks on Related Articles, Entrez will indicate that there are 101 neighbours associated with the original Cayatte reference. The first reference in the list is the same Cayatte paper because, by definition, it is most related to itself (the "parent"). The order in which the neighboured entries follows is from most statistically similar downward. Thus, the entry closest to the parent is deemed to be the closest in subject matter to the parent. By scanning the titles, the user can easily find related information on other studies that look at the pharmacology of aspirin in atherosclerosis as well as quickly amass a bibliography of relevant references. This is a particularly useful and time-saving functionality when one is writing grants or papers because abstracts can be scanned and papers of real interest identified before one heads off for the library stacks.

The next link in the series is labeled Books, and clicking on that link will take the user to a heavily hyperlinked version of the original citation. The highlighted words in this view correspond to keywords that can take the user to full-text books that are available through NCBI. The first of these books to be made available is *Molecular Biology of the Cell*. Following the Cayette example, if the user clicks on atherosclerosis at this point, it will take them to the relevant part of the textbook, a section devoted to how cells import cholesterol by receptor-mediated endocytosis. From this page, the user can navigate through this particular unit, gathering more general information on transport from the plasma membrane via endosomes and vesicular traffic in the secretory and endocytic pathways.

The final link in the series in the upper right is LinkOut. This feature provides a list of third-party Web sites and resources relating to the Entrez entry being viewed, such as full-text of articles that can be displayed directly through the Web browser, or the capability of ordering the document through services such as Loansome Doc. A "cubby" service for LinkOut enables users to customize which links are displayed in a LinkOut view. Another way of getting to the full text of an article is by following a direct link to the publisher's Web site. In the Abstract view for the Cayette example, a button directly under the citation is marked ATVB, for *Arteriosclerosis*, *Thrombosis*, and *Vascular Biology*, the journal in which the paper is published. With the proper individual or institutional subscription privileges, one would be able to immediately see the full text of the paper, including all figures and tables.

There is another way to perform an Entrez query, involving some built-in features of the system. Consider an example in which one is attempting to find all genes coding for DNA-binding proteins in methanobacteria. In this case, the search would begin by setting the Search pull-down menu to Nucleotide and then typing the term DNA-binding into the text box. This search returns 23,797 entries in which the search term appears. At this point, to narrow down the search, the user would click on the Limits hyperlink, directly below the text box. This brings the user to a new page that allows the search to be limited, as implied by the name of the hyperlink. Here, the search will be limited by organism, so the Limited To pulldown is changed to Organism, and the word methanobacterium is typed into the search box Clicking Go will now return all of the entries in which *Methanobacterium* is the organism. The results from the first search can also be combined with those from the second by clinking on the History hyperlink below the text box, resulting in a list of recent queries. The list shows the individual queries, whether those queries were field-limited, the time at which the query was performed, and how many entries that individual query returned. To combine two separate queries into one, the user simply combines the queries by number; in this case, because the queries are numbered #8 and #9, the syntax would be #8 AND #9. Clicking Preview regenerates a table, showing the new, combined query as #10, containing three entries. Alternatively, clicking Go shows the user the three entries, in the now-familiar nucleotide summary format.

As before, there are a series of hyperlinks to the upper right of each entry; three are shown for the first entry, which is for the *M. thermoautotrophicum* tfx gene. The PubMed link takes the user back to the bibliographic entry or entries corresponding to this GenBank entry. Clicking on Protein brings into play one of the hard link relationships, showing the GenPept entry that corresponds to the tfx gene's conceptual translation. Notice that, within the entry itself, the scientific name of the organism is represented as hypertext; clicking on that link takes the user to the NCBI Taxonomy database, where information on that particular organism's lineage is available. One of the useful views at this level is the Graphics view; this view attempts to show graphically all of the features described within the entry's feature table, providing a very useful overview, particularly when the feature table is very long. The Related Sequences link shows all sequences similar to that of the tfx gene at the nucleotide level, in essence showing the results of a precomputed BLAST search.

The last part of Entrez to be discussed deals with structures. Structure queries can be done directly by specifying Structure in the Search pull-down menu. For example, suppose that one wishes to find out information about the structure of HMG-box B from rat, whose accession number is 1HMF. Typing 1HMF into the query box leads the user to the structure summary page for 1HMF, which has a decidedly different format than any of the pages seen so far. This page shows details from the header of the source MMDB document (which is derived from PDB), links to PubMed and to the taxonomy of the source organism, and links to both sequence and structure neighbours. The Sequence Neighbours links show neighbours to 1HMF on the basis of sequence—that is, by BLAST search—thus, although this is a *structure* entry, it is important to realize that sequence neighbours have nothing to do with the structural information, at least not directly. To get information about related structures, one of the Structure Neighbour links can be followed, producing a table of neighbours as assessed by VAST. For a user interested in gleaning initial impressions about the shape of a protein, the Cn3D plug-in, invoked by clicking on View/ Save Structure, provides a powerful interface, giving far more information than anyone could deduce from simply examining a string of letters (the sequence of the protein).

The protein may be rotated along its axes by means of the scroll bars on the bottom, top, and right-hand side of the window or may be freely rotated by clicking and holding down the mouse key while the cursor is within the structure window and then dragging. Users are able to zoom in on particular parts of the structure or change the colouration of the figure, to determine specific structural features about the protein. For instance, Spacefilling and Hydrophobicity were chosen as the Render and Colour options, respectively. In addition, users can save coordinate information to a file and view the data using third-party applications such as Kinemage and RasMol.

Finally, at any point along the way in using Entrez, if there are partial or complete search results that the user wishes to retain while moving onto a new query, the Add to Clipboard button can be pushed. This stores the results of the current query, which the user can return to by clicking the Clipboard hyperlink directly under the text box. The clipboard holds a maximum of 500 items, and the items are held in memory for 1 h.

LocusLink

The Entrez system revolves necessarily around the individual entries making up the various component databases that are part of the Entrez search space. Another way to think about this search space is to organize it around discrete genetic loci. NCBI LocusLink does just this, providing a single query interface to various types of information regarding a given genetic locus, such as phenotypes, map locations, and homologies to other genes. The LocusLink search space currently includes information from humans, mice, rats, fruit flies, and zebrafish.

With the use of the gene for the high-mobility group protein HMG1 as an example, the LocusLink query begins by the user simply typing the name of the gene into the query box appearing at the top of the LocusLink home page. Alternatively, the user could select the gene of interest from an alphabetical list. The query on HMG1 returns three LocusLink entries, from human, mouse, and rat. In this view, the user is given the Locus ID in the first column; the Locus ID is intended to be a unique, stable identifier that is associated with this gene locus. Clicking on the Locus ID for the human (3146) produces the LocusLink Report view. The Report view begins with links to external sources of information, shown as coloured buttons at the top of the page. In this particular report, the links would lead the user to PubMed (Pub), UniGene (UG), the dbSNP variation database (VAR), HomoloGene (HOMOL), and the Genome Database (GDB). These offsite links will change from entry to entry, depending on what information is available for the gene of interest.

Continuing down the Report view, in the section marked Locus Information, the user is presented with the official gene symbol, along with any alternate symbols that may have traditionally been used in the literature or in sequence records. This section would also include information on the name of the gene product, any aliases for the name of the gene product, the Enzyme Commission number, the name of any diseases that result from variants at this gene locus, and links to OMIM and UniGene. Only those pieces of information that are known or are applicable to this particular gene locus are shown.

In the section labeled Map Information, the report shows what chromosome this locus is on, the cytogenetic and genetic map positions, when known, and any STS markers that are found within the mRNA corresponding to this locus. There is a hyperlink that can take the user to the Entrez Map Viewer, showing the position of this locus and the relationship of this locus to surrounding loci. The Map Viewer shows the chromosomal ideogram to the left, with the highlighted region marked by a thick bar to the right of the ideogram. The user can zoom in or out by clicking on the icon above the ideogram. In the main window, the user is presented with both the cytogenetic and sequence map. In this particular view, 20 genes are shown, with the original locus of interest highlighted. As with most graphical views of this type, the majority of the elements in this view are clickable, taking the user to more information about that particular part of either the cytogenetic or sequence map.

RefSeq is short for the NCBI Reference Sequence Project, which is an effort aimed at providing a stable, reference sequence for all of the molecules corresponding to the central dogma of biology. The intention is that these reference sequences will provide a framework for making functional annotations, as well as for information regarding mutations, gene expression, and polymorphisms. The sequences listed in this section represent the sequences that were used to build the corresponding RefSeq record. Notice that RefSeq nucleotide records are preceded by NM and that protein records are preceded by NP. A blue button appears next to protein information if there is also structural information available about the protein. The final portion of the LocusLink report shows the GenBank accession numbers that correspond to this locus. The middle column indicates the molecule type for these GenBank entries, with m standing for mRNA, g for genomic DNA, e for an EST, and u for undetermined. In this particular case, there is also a link to the GeneCard for HMG1; clicking on that hyperlink takes the user to the GeneCards database at the Weizmann Institute, providing a concise summary of the function of this gene.

Another way to proceed through the information is to return to the query and use the linked alphabet blocks shown to the right of each of the entries. In turn, these links are

- P, for PubMed bibliographic entries corresponding to the locus;
- O, for the Online Mendelian Inheritance in Man summary on this locus;
- R, for the RefSeq entries corresponding to the locus;
- G, for the individual GenBank entries related to the locus; these entries will correspond to the RefSeq entry, as shown in the LocusLink Report view;
- H, for HomoloGene. HomoloGene allows the user to find both orthologs and homologs for genes represented in LocusLink and UniGene, the assignments being made based on sequence similarity;
- U, whether this locus is part of a UniGene cluster; and
- V, for variation data on this locus contained within dbSNP.

When following either the PubMed or GenBank links, the user is, in essence, returned to the Entrez search space, enabling the user to take advantage of Entrez's navigational features once again.

Sequence Databases beyond NCBI

Although it may appear from this discussion that NCBI is the center of the sequence universe, many specialized sequence databases throughout the world serve specific groups in the scientific community. Often, these databases provide additional information such as phenotypes, experimental conditions, strain crosses, and map features. The data are of great importance to these subsets of the scientific community, inasmuch as they can influence rational experimental design, but such types of data do not always fit neatly within the confines of the NCBI data model. Development of specialized databases necessarily ensued, but they are intended to be used as an adjunct to GenBank, not in place of it. It is impossible to discuss all of these kinds of databases here, but, to emphasize the sheer number of such databases that exist, *Nucleic Acids Research* devotes its first issue every year to papers describing these databases.

An example of a specialized organismal database is the *Saccharomyces* Genome Database (SGD), housed at the Stanford Human Genome Center. The database provides a very simple search interface that allows text-based searches by gene name, gene information, clone, protein information, sequence name, author name, or full text. For example, using Gene Name as the search topic and hho1 as the name of the gene to be searched for produces a SacchDB information window showing all known information on locus HHO1. This window provides jumping-off points to other databases, such as GenBank/GenPept, MIPS, and the Yeast Protein Database (YPD). Following the link to Sacch3D for this entry provides information on structural homologs of the HHO1 protein product found in PDB, links to secondary and tertiary structure prediction sites, and precomputed BLAST reports against a number of query databases. Returning to the Locus window and clicking on the map in the upper right-hand corner, the user finds a graphical view of the area surrounding the locus in question. Available views include physical maps, genetic maps, and chromosomal physical maps, among others. Note the thick bar at the top of the figure, which gives the position of the current view with respect to the centromere. Clicking on that bar allows the user to move along the chromosome, and clicking on individual gene or ORF name (or, as the authors cite in the figure legend, "any little colourful bar") gives more detailed information about that particular region.

Another example of an organism-specific database is FlyBase, whose goal is to maintain comprehensive information on the genetics and molecular biology of *Drosophila*. The information found in FlyBase includes an extensive *Drosophila* bibliography, addresses of over 6,000 researchers involved in *Drosophila* projects, a compilation of information on over 51,500 alleles of more than 13,200 genes, information about the expression and properties of over 4,800 transcripts and 2,500 proteins, and descriptions of over 16,700 chromosomal aberrations. Also included is relevant mapping information, functional information on gene products, lists of stock centers and genomic clones, and information from allied databases. Searches on any of these "fields" can be done through a simple search mechanism. For example, searching by gene symbol using capu as the search term brings up a record for a gene named cappuccino, which is required for the proper polarity of the developing *Drosophila* oocyte. Calling up the cytogenetic map view generates a map showing the gene and cytologic location of cappuccino and other genes in that immediate area, and users can click on any of the gene bars to bring up detailed information on that particular gene. The view can be changed by selecting one of the Class buttons at the bottom of the window, so that a graphical view of cosmids, deficiencies, duplications, inversions, transpositions, translocations, or other aberrations can be examined instead.

Medical Databases

Although the focus of this chapter (and the book in general) is on sequences, databases cataloguing and organizing sequence information are not the only kinds of databases useful to the biologist. An example of such a non-sequence-based information resource that is tremendously useful in genomics is Online Mendelian Inheritance in Man (OMIM), the electronic version of the catalog of human genes and genetic disorders founded by Victor McKusick at The Johns Hopkins University. OMIM provides concise textual information from the published literature on most human conditions having a genetic basis, as well as pictures illustrating the condition or disorder (where appropriate) and full citation information. Because the online version of OMIM is housed at NCBI, links to Entrez are provided from all references cited within each OMIM entry.

OMIM has a defined numbering system in which each entry is assigned a unique number, similar to an accession number, but certain positions within that number indicate information about the genetic disorder itself. For example, the first digit represents the mode of inheritance of the disorder: 1 stands for autosomal dominant, 2 for autosomal recessive, 3 for X-linked locus or phenotype, 4 for Y-linked locus or phenotype, 5 for mitochondrial, and 6 for autosomal locus or phenotype. (The distinction between 1 or 2 and 6 is that entries catalogued before May 1994 were assigned either a 1 or 2, whereas entries after that date were assigned a 6 regardless of whether the mode of inheritance was dominant or recessive.) An asterisk preceding a number indicates that the phenotype caused by the gene at this locus is *not* influenced by genes at other loci; however, the disorder itself may be caused by mutations at multiple loci. Disorders for which no mode of inheritance has been determined do not carry asterisks. Finally, a pound sign (#) indicates that the phenotype is caused by two or more genetic mutations.

OMIM searches are very easy to perform. The search engine performs a simple query based on one or more words typed into a search window. A list of documents containing the query words is returned, and users can select one or more disorders from this list to look at the full text of the OMIM entry. The entries include information such as the gene symbol, alternate names for the disease, a description of the disease (including clinical, biochemical, and cytogenetic features), details on the mode of inheritance (including mapping information), a clinical synopsis, and references. A particularly useful feature is lists of allelic variants; a short description is given after each allelic variant of the clinical or biochemical outcome of that particular mutation. There are currently over 1,000 gene entries containing at least one allelic variant that either causes or is associated with a discrete phenotype in humans.

Sequence Alignment and Database Searching

There is a long tradition in biology of comparative analysis leading to discovery. For instance, Darwin's comparison of morphological features of the Galapagos finches and other species ultimately led him to postulate the theory of natural selection. In essence, we are performing the same type of analysis today when we compare the sequences of genes and proteins but in much greater detail. In this activity, the similarities and differences—at the level of individual bases or amino acids—are analyzed, with the aim of inferring structural, functional, and evolutionary relationships among the sequences under study. The most common comparative method is *sequence alignment*, which provides an explicit mapping between the residues of two or more sequences. The number of sequences available for comparison has grown explosively since the 1970s, when development of rapid DNA sequencing methodology sparked the "big bang" of sequence information expansion. Comparison of one sequence to the entire database of known sequences is an important discovery technique that should be at the disposal of all molecular biologists. Over the past 30 years, improvements in the speed and sophistication of sequence alignment algorithms, not to mention performance of computers, have more than kept pace with the growth in the size of the sequence databases. Today, with the complete genomes and large cDNA sequence collections available for many organisms, we are in the era of "comparative genomics," in which the full gene complement of two organisms can be compared with one another.

Evolutionary Basis of Sequence Alignment

One goal of sequence alignment is to enable the researcher to determine whether two sequences display sufficient similarity such that an inference of homology is justified. Although these two terms are often interchanged in popular usage, let us distinguish them to avoid confusion in the current discussion. *Similarity* is an observable quantity that might be expressed as, say, percent identity or some other suitable measure. *Homology*, on the other hand, refers to a conclusion drawn from these data that two genes share a common evolutionary history. Genes either are or are not homologous—there are no degrees for homology as there are for similarity. For example, an alignment between the homologous trypsin proteins from *Mus musculus* (house mouse) and *Astracus astracus* (broad-fingered crayfish), from which it can be calculated that these two sequences have 41% identity.

Bearing in mind the goal of inferring evolutionary relationships, it is fitting that most alignment methods try, at least to some extent, to model the molecular mechanisms by which sequences evolve. Although it is presumed that homologous sequences have diverged from a common ancestral sequence through iterative molecular changes, it is actually known what the ancestral sequence was (barring the possibility that DNA could be recovered from a fossil); all that can be observed are

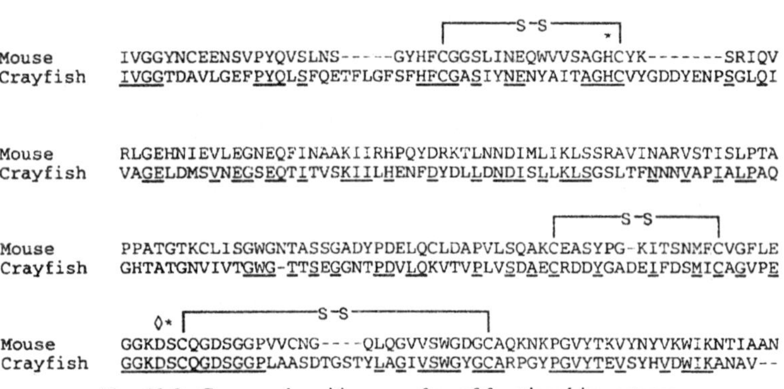

Fig. 10.2. Conserved position are often of functional importance.

the raw sequences from extant organisms. The changes that occur during divergence from the common ancestor can be categorized as substitutions, insertions, and deletions. In the ideal case, in which a sequence alignment genuinely reflects the evolutionary history of two genes or proteins, residues that have been aligned but are not identical would represent substitutions. Regions where the residues of one sequence correspond to nothing in the other would be interpreted as either an insertion into one sequence or a deletion from the other. These *gaps* are usually represented in the alignment as consecutive dashes (or other punctuation character) aligned with letters.

In a residue-by-residue alignment, it is often apparent that certain regions of a protein, or perhaps specific amino acids, are more highly conserved than others. This information may be suggestive as to which residues are most crucial for maintaining a protein's structure or function. In the trypsin alignment, the active site residues that determine substrate specificity and provide the "charge relay system" of serine proteases correspond to conserved positions, as do the cysteines residues that form several disulfide bonds important for maintaining the enzyme's structure. On the other hand, there may be other positions that do not play a significant functional role yet happen to be identical for historical reasons. Particular caution should be taken when the sequences are taken from very closely related species because similarities may be more reflective of history than of function. For example, regions of high sequence similarity between mouse and rat homologs may simply be those that have not had sufficient time to diverge. Nevertheless, sequence alignments provide a useful way to gain new insights by leveraging existing knowledge, such as deducing structural and functional properties of a novel protein from comparisons to those that have been well studied. It must be emphasized, however, that these inferences should always be tested experimentally and *not* assumed to be correct based on computational analysis alone.

By observing a surprisingly high degree of sequence similarity between two genes or proteins, we might infer that they share a common evolutionary history, and from this it might be anticipated that they should also have similar biological functions. But again, this should be treated as hypothetical until tested experimentally. Zeta-crystallin, for instance, is a component of the transparent lens matrix of the vertebrate eye. However, on the basis of extended sequence similarity, it can be inferred that its homolog in *E. coli* is the metabolic enzyme quinone oxidoreductase. Despite the common ancestry, the function has changed

```
Human-ZCr    MATGQKLMRAVRVFEFGGPEVLKLRSDIAVPIPKDHQVLIKVHACGVNPVETYIRSGTYS
Ecoli-QOR    ------MATRIEFHKHGGPEVLQA-VEFTPADPAENEIQVENKAIGINFIDTYIRSGLYP
                   .   .     ******.   . .   * .... .  .* *.* ..****** *

Human-ZCr    RKPLLPYTPGSDVAGVIEAVGDNASAFKKGDRVFTSSTISGGYAEYALAADHTVYKLPEK
Ecoli-QOR    -PPSLPSGLGTEAAGIVSKVGSGVKHIKAGDRVVYAQSALGAYSSVHNIIADKAAILPAA
               * **   *.. **..  ** .    * ****  . .  * *.            **

Human-ZCr    LDFKQGAAIGIPYFTAYRALIHSACVKAGESVLVHGASGGVGLAACQIARAYGLKILGTA
Ecoli-QOR    ISFEQAAASFLKGLTVYYLLRKTYEIKPDEQFLFHAAAGGVGLIACQWAKALGAKLIGTV
             . * * **  .   * *  * ..  .*  *  * * *.***** *** *.* * *..**

Human-ZCr    GTEEGQKIVLQNGAHEVFNHREVNYIDKIKKYVGEKGIDIIIEMLANVNLSKDLSLLSHG
Ecoli-QOR    GTAQKAQSALKAGAWQVINYREEDLVERLKEITGGKKVRVVYDSVGRDTWERSLDCLQRR
             ** .  .  *. ** .* * **   ....*   * * . .. . .   .  . *  * .

Human-ZCr    GRVIVVG-SRGTIEINPRDTMAKES----SIIGVTLFSSTKEEFQQYAAALQAGMEIGWL
Ecoli-QOR    GLMVSFGNSSGAVTGVNLGILNQKGSLYVTRPSLQGYITTREELTEASNELFSLIASGVI
             * ..  * * *..       . .      .   .  . .*.**  . .  * . .  * .

Human-ZCr    KPVIGSQ--YPLEKVAEAHENIIHGSGATGKMILLL
Ecoli-QOR    KVDVAEQQKYPLKDAQRAHE-ILESRATQGSSLLIP
             *  .  *  ***     *** *.    . *  .*.
```

Fig. 10.3. Optimal global sequence alignment.

during evolution. This is analogous to a railroad car that has been converted into a diner: inspection of the exterior structure reveals the structure's history, but relying exclusively on this information may lead to an erroneous conclusion about its current function. When a gene adapts to a new niche, it might also be anticipated that the pattern of conserved positions would change. For example, active site residues should be conserved so long as the protein plays a role in catalysis but could drift once the protein takes on a different function.

The earliest sequence alignment methods were applicable to a simple type of relationship in which the sequences show easily detectable similarity along their entire lengths. An alignment that essentially spans the full extents of the input sequences is called a *global alignment*. The trypsin and quinone oxidoreductase/zetacrystallin alignments discussed above are both examples of global alignments. Proteins consisting of a single globular domain can often be aligned using a global strategy as can any homologous sequences that have not diverged substantially.

Modular Nature of Proteins

Many proteins do not display global patterns of similarity but instead appear to be mosaics of modular domains. One example of this shows the modular structure of two proteins involved in blood clotting: coagulation factor XII (F12) and tissue-type plasminogen activator (PLAT). Besides the catalytic domain, which provides the serine protease activity, these proteins have different numbers of other structural modules: two types of fibronectin repeats, a domain with similarity to epidermal growth factor, and a module that is called a "kringle" domain. These modules can be repeated or appear in different orders. Patterns of modularity often arises by in-frame exchange of whole exons. Global alignment methods do not take this phenomenon into account, which is understandable considering that they were developed before the exon/intron structure of genes had been discovered. In most cases, it is advisable to instead use a sequence comparison method that can produce a *local alignment*. Such an alignment consists of paired subsequences that may be surrounded by residues that are completely unrelated. Consequently, users should bear in mind that some local similarities could be missed if a global alignment strategy is applied inappropriately. Another obvious case in which local alignments are desired is the alignment of the nucleotide sequence of a spliced mRNA to its genomic sequence, where each exon would be a distinct local alignment.

Dot-matrix representations have enjoyed a widespread popularity, in part because of their ability to reveal complex relationships involving multiple regions of local similarity. An example of this approach is

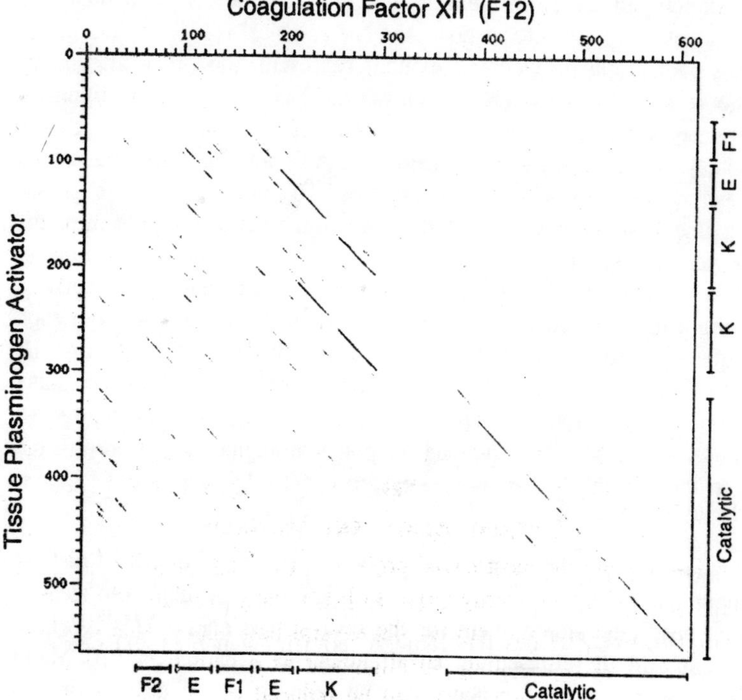

Fig. 10.4. Dot matrix sequence comparison.

shown in Figure 10.4, in which the F12 and PLAT protein sequences have been compared using dotter. The basic idea is to use the sequences as the coordinates of a two-dimensional graph and then plot points of correspondence within its interior. Each dot usually indicates that, within some small window, the sequence similarity is above some cutoff (or a range of cutoffs with the use of dotter, each plotted using a different shade of gray). When two sequences are consistently matching over an extended region, the dots will merge to form a diagonal line segment. It is instructive to compare the positions of the diagonals in dot-matrix with the known modular structure of the two proteins. In particular, note the way in which repeated domains appear: starting with the kringle domain in the PLAT and scanning horizontally, two diagonal segments may be seen, corresponding to the two kringle domains present in the F12 sequence. Although more sophisticated methods for finding local similarities are now available, dot-matrix representations have remained popular as illustrative tools.

In a dot-matrix representation, certain patterns of dots may appear to sketch out a "path," but it is up to the viewer to deduce the alignment from this information. Another graphical representation known as a path graph provides an explicit representation of an alignment. The relationship between the dot-matrix, path graph, and alignment representations for the EGF similarity domain present in both the tissue-type plasminogen activator (PLAT) and the urokinase-type plasminogen activator (PLAU) proteins. To understand a path graph, imagine a twodimensional lattice in which the vertices represent points between the sequence residues (as opposed to the residues themselves, as in the case of the dot-matrix). An edge that connects two vertices along a diagonal corresponds to the pairing of one residue from each sequence. Horizontal and vertical edges pair a residue from one sequence with nothing in the other; in other words, these edges constitute a gap in the alignment. The entire graph corresponds to the *search space*, which must be examined for potential alignments. Each possible path through this space corresponds to exactly one alignment.

Optimal Alignment Methods

For any but the most trivial problems, the total number of distinct alignments is extraordinarily large, so it is usually of interest to identify the "best" one among them (or the several best ones). This is where the concept of representing an alignment as a path pays off. Many problems in computer science can be reduced to the task of finding the optimal path through a graph, and efficient algorithms have been

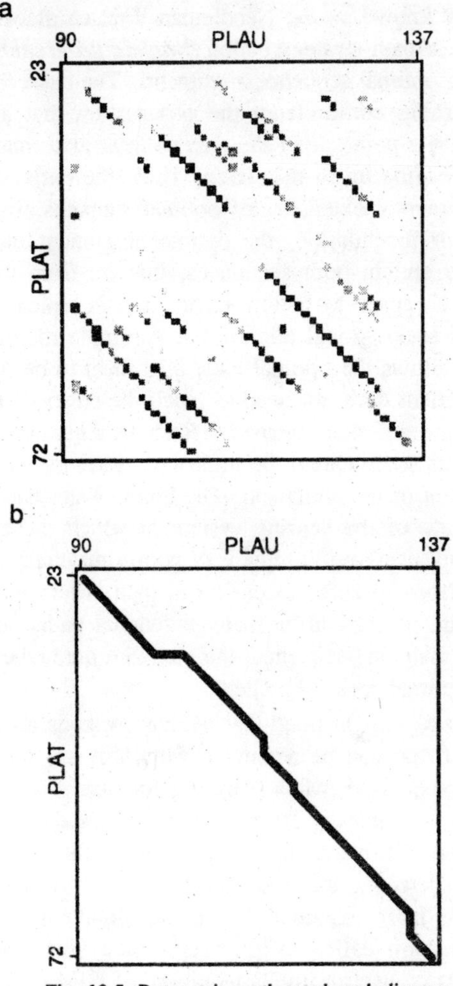

Fig. 10.5. Dot-matrix, path graph and alignment.

developed for this purpose. One requirement is a means of assigning a quality score to each possible path (alignment). Normally, this is accomplished by summing the incremental contributions of each step along its route. More sophisticated scoring schemes are discussed below, but for now let us assume that some positive incremental scores will be used for aligning identical residues, with negative scores used for substitutions and gaps. According to this definition of alignment quality, finding the path whose total score is maximal will give us the best sequence alignment.

What is today known as the Needleman-Wunsch algorithm is an application of a best-path strategy called *dynamic programming* to the problem of finding optimal sequence alignments. The basic idea behind dynamic programming comes from the observation that any partial subpath that ends at a point along the true optimal path must itself be the optimal path leading up to that point. Thus, the optimal path can be found by incremental extension of optimal subpaths. In the basic Needleman-Wunsch formulation, the optimal alignment must extend from beginning to end in both sequences, that is, from the top-left corner in the search space to bottom-right (as it is typically drawn). In other words, it seeks global alignments. A simple modification to the basic strategy allows the optimal local alignment to be found. The path for this alignment does not need to reach the edges of the search graph but may begin and end internally. Such an alignment would be locally optimal if its score cannot be improved either by increasing or decreasing the extent of the alignment. The Smith-Waterman algorithm relies on a property of the scoring system in which the cumulative score for a path will decrease in regions of poorly matching sequences. When the score drops to zero, extension of path is terminated and a new one can begin. There can be many individual paths bounded by regions of poorly matching sequence; the one with the highest score is reported as the optimal local alignment.

It is important to bear in mind that optimal methods always report the best alignment that can be achieved, even if it has no biological meaning. On the other hand, when searching for local alignments there may be several significant alignments, so it is a mistake to look only at the optimal one. Refinements to the Smith-Waterman algorithm were proposed for detecting the *k* best nonintersecting local alignments. These ideas were later extended in the development of the SIM algorithm. A program called lalign (distributed with the FASTA package) provides a useful implementation of SIM. Looking for suboptimal alignments is especially important when comparing multimodule proteins. The second and third alignments represent functional modules that would have been missed by a standard Smith-Waterman search, which would have reported only the first (optimal) alignment.

Substitution Scores and Gap Penalties

The scoring system described above made use of a simple match/ mismatch scheme, but, when comparing proteins, we can increase sensitivity to weak alignments through the use of a *substitution matrix*.

```
Comparison of:
(A) f9-human.aa >F9  gi|119772|sp|P00740|FA9_HUMAN COAGULATION FA  - 461 aa
(B) f12-hum.aa >F12   gi|119763|sp|P00748|FA12_HUMAN COAGULATION  - 615 aa
using protein matrix

①    35.4% identity in 254 aa overlap; score:  358

         220       230       240       250       260       270
F9    QSFNDFTRVVGGEDAKPGQFPWQVVLNGKVDAFCGGSIVNEKWIVTAAHCVE---TGVKI
      .:....::::::  : .:. :. ..:     ..::.::...  :..::::::..   .  ..
F12   KSLSSMTRVVGGLVALRGAHPYIAALY-WGHSFCAGSLIAPCWVLTAAHCLQDRPAPEDL
         370       380       390       400       410       420

         280       290       300       310       320       330
F9    TVVAGEHNIEETEHTEQKRNVIRIIPHHNYNAAINKYNHDIALLELDEPL-----VLNSY
      ::: :... ... .. :. .: .   :......  .:.::.::: :.:       .:..:
F12   TVVLGQERRNHSCEPCQTLAVRSYRLHEAFSPV--SYQHDLALLRLQEDADGSCALLSPY
         430       440       450       460       470       480

           340       350       360       370       380
F9    VTPICIADKEYTNIFLKFGSGYVSGWGRVFHKGRS-ALVLQYLRVPLVDRATCLRSTKF-
      : :.:... .       ..    :.:::. :. . . :   ::   .::... .  :  ..
F12   VQPVCLPSGAARPSETTLCQ--VAGWGHQFEGAEEYASFLQEAQVPFLSLERCSAPDVHG
           490       500       510       520       530

      390       400       410       420       430       440
F9    -TIYNNMFCAGFHEGGRDSCQGDSGGPHVTEVEGTS---FLTGIISWGEECAMKGKYGIY
       .:  .:.:::: ::: :.::::::::: : : ....    : ::::::..:.  ..: :.:
F12   SSILPGMLCAGFLEGGTDACQGDSGGPLVCEDQAAERRLTLQGIISWGSGCGDRNKPGVY
    540       550       560       570       580       590

        450
F9    TKVSRYVNWIKEKT
      :.:. :..::.:.:
F12   TDVAYYLAWIREHT
    600       610

----------

②    34.7% identity in 49 aa overlap; score:  120

           100       110       120       130       140
F9    VDGDQCESNPCLNGGSCKDDINSYECWCPFGFEGKNCELDVTCNIKNGR
      .....: .::::.::.: .  .   : :: :..:  :..:.. .  .::
F12   LASQACRTNPCLHGGRCLEVEGHRLCHCPVGYTGPFCDVDTKASCYDGR
          180       190       200       210       220

----------

③    33.3% identity in 36 aa overlap; score:   87

        100       110       120
F9    DQCESN-PCLNGGSCKDDINSYECWCPFGFEGKNCE
      :.:... :: .::.: .  .. .: ::  ..:..:.
F12   DHCSKHSPCQKGGTCVNMPSGPHCLCPQHLTGNHCQ
        100       110       120       130

----------
```

Fig. 10.6. Optimal and suboptimal local alignment. The three best alignments found when using lalignto align the sequences of human coagulation factor IX and coagulation factor XII.

It is well known that certain amino acids can substitute easily for one another in related proteins, presumably because of their similar physicochemical properties. Examples of these "conservative substitutions" include isoleucine for valine (both small and hydrophobic) and serine for threonine (both polar). When calculating alignment scores, identical amino acids should be given greater value than substitutions, but conservative substitutions should also be greater than nonconservative changes. In other words, a range of values is desired. Furthermore, different sets of values may be desired for comparing very similar sequences (e.g., a mouse gene and its rat homolog) as opposed to highly divergent sequences (e.g., mouse and yeast genes). These considerations can be dealt with in a flexible manner through the use of a substitution matrix, in which the score for any pair of amino acids can be easily looked up.

The first substitution matrices to gain widespread usage were those based on the point accepted mutation (PAM) model of evolution. One PAM is a unit of evolutionary divergence in which 1% of the amino acids have been changed. This does not imply that after 100 PAMs every amino acid will be different; some positions may change several times, perhaps even reverting to the original amino acid, whereas others may not change at all. If there were no selection for fitness, the frequencies of each possible substitution would be primarily influenced by the overall frequencies of the different amino acids (called the *background frequencies*). However, in related proteins, the observed substitution frequencies (called the *target frequencies*) are biased toward those that do not seriously disrupt the protein's function. In other words, these are point mutations that have been "accepted" during evolution. Dayhoff and coworkers were the first to explicitly use a *log-odds* approach, in which the substitution scores in the matrix are proportional to the natural log of the ratio of target frequencies to background frequencies. To estimate the target frequencies, pairs of very closely related sequences (which could be aligned unambiguously without the aid of a substitution matrix) were used to collect mutation frequencies corresponding to 1 PAM, and these data were then extrapolated to a distance of 250 PAMs. Although PAM250 was the only matrix published by Dayhoff et al. (1978), the underlying mutation data can be extrapolated to other PAM distances to produce a family of matrices. When aligning sequences that are highly divergent, best results are obtained at higher PAM values, such as PAM200 or PAM250. Matrices constructed from lower PAM values can be used if the sequences have a greater degree of similarity.

The BLOSUM substitution matrices have been constructed in a similar fashion, but make use of a different strategy for estimating the target frequencies. The underlying data are derived from the BLOCKS database, which contains local multiple alignments ("blocks") involving distantly related sequences (as opposed to the closely related sequences used for PAM). Although there is no evolutionary model in this case, it is advantageous to have data generated by direct observation, rather than extrapolation. As with the PAM model, there is a numbered series of BLOSUM matrices, but the number in this case refers to the maximum level of identity that sequences may have and still contribute independently to the model. For example, with the BLOSUM62 matrix, sequences having at least 62% identity are merged into a single sequence, so that the substitution frequencies are more heavily influenced by sequences that are more divergent than this cutoff. Substitution matrices have been constructed using higher cutoffs (up to BLOSUM90) for comparing very similar sequences and lower cutoffs (down to BLOSUM30) for highly divergent sequences.

It is desirable to allow some gaps to be introduced into an alignment to compensate for insertions and deletions but not so many that the alignment asserts an implausible series of molecular alterations. This is accomplished by deducting some amount from the alignment score for each gap introduced. Although a number of strategies have been proposed for penalizing gaps, the most common formulation, known as *affine gap penalties*, involves a fixed deduction for introducing a gap plus an additional deduction proportional to the length of the gap. This is governed by two parameters: G, sometimes called the gap-opening penalty, and L, the gapextension penalty. For a gap of length n, the total deduction would be $G + Ln$. Unfortunately, the selection of gap parameters is highly empirical; there is little theory to support the choice of any particular set of values. However, it is common to use a high value for G (around 10–15, in the context of BLOSUM62) and a low value for L (around 1 or 2). The rationale for this is that insertion and mutation events are rare, but, when they do occur, several adjacent residues may be involved.

Statistical Significance of Alignments

For any given alignment, one can calculate a score representing the quality of the alignment, but an important question is whether or not this score is high enough to provide evidence of homology. In addressing this question, it is helpful to have some notion of how high of a score can be expected due purely to chance alone. Unfortunately, there is no mathematical theory to describe the expected distribution of scores for global alignments. One of the few methods available for assessing their significance is to compare the observed alignment score with those of many alignments made from random sequences of the same length and composition as those under study.

However, for local alignments, the situation is much better. A statistical model advanced by Karlin and Altschul provides a mathematical theory to describe the expected distribution of random local alignment scores. The form of the probability density function is known as the *extreme value distribution*. This is worth noting because application of the more familiar normal distribution can result in greatly exaggerated claims of significance. The extreme value distribution is characterized by two parameters, K and λ, which should be tailored for the particular set of alignment scoring rules and residue background frequencies at hand. Although analytical calculation of these parameters can currently be done only for alignments that lack gaps, methods have been developed to estimate appropriate values of K and λ for gapped alignments. By relating an observed alignment score S to the expected distribution, it is possible to calculate statistical significance in the form of an *E value*. The simple interpretation of an E value is the number of alignments with scores at least equal to S that would be expected by chance alone. The significance of an alignment also depends on the size of the search space that was used; larger databases produce more chance alignments. The search space has typically been calculated as the product of the sequence lengths, but, for correct statistics, the lengths must be reduced by the expected length of a local alignment to avoid an "edge effect". This is due to the fact that an alignment that begins near the edge of the search space will run out of sequence before it can achieve a significant score.

Database Similarity Searching

The discussion so far has focused on the alignment of specific pairs of sequences, but, for a newly determined sequence, one would generally have no way of knowing the appropriate sequence (or sequences) to use in such a comparison. Database similarity searching allows us to determine which of the hundreds of thousands of sequences present in the database are potentially related to a particular sequence of interest. This process sometimes leads to unexpected discoveries. The first "eureka moment" with this strategy came when the viral oncogene v-*sis* was found to be a modified form of the normal cellular gene that encodes platelet-derived growth factor. At the time of this discovery, sequence databases were small enough that such a finding might have been considered surprising. Today, however, it would be much more surprising to perform a database search and *not* get a hit. Large numbers of partial sequences representing novel human and mouse genes have been deposited in GenBank as a result of a number of expressed sequence tag projects. The genomes of *S. cerevisiae*, *C. elegans*, and *D. melanogaster* have been completely sequenced, as have many bacterial and viral genomes. More recently, an intermediate form of the human genome is now available in the form of a "working draft." Given this explosion of data, the new challenge has become how to focus searches in such a way as to reduce search times and limit the number of results that must be examined.

In database searching, the basic operation is to sequentially align a *query sequence* to each *subject sequence* in the database. The results are reported as a ranked *hit list* followed by a series of individual sequence alignments, plus various scores and statistics. As will be discussed in more detail below, that choice of search program, sequence database, and various optional parameters can have an impact on the effectiveness of a search. Furthermore, there are various interfaces to these facilities such as console-style commands, Web-based forms, and E-mail. One advantage of this approach is that, for any interesting alignment observed, complete annotation and literature citations can be obtained simply by following hypertext links to the original sequences entries and related on-line literature.

Current sequence databases are immense and have continued to increase at an exponential rate, making straightforward application of dynamic programming methods impractical for database searching. One solution is to use massively parallel computers and other specialized hardware, but, for the purposes of this discussion, we will consider only what can be done using general-purpose computers. With optimal methods being impractical, it is necessary to resort to heuristic methods, which make use of approximations to significantly speed up sequence comparisons, but with a small risk that true alignments can be missed. One heuristic method is based on the strategy of breaking a sequence up into short runs of consecutive letters called *words*. Word-based methods were introduced in the early 1980s and are used by virtually all popular search programs in use today. The basic idea is that an alignment representing a true sequence relationship will contain at least one word that is common to both sequences. These *word hits* can be identified extremely rapidly by preindexing all words from the query and then consulting the index as the database is scanned.

FASTA

The first widely-used program for database similarity searching was FASTA. To achieve a high degree of sensitivity, this program performs optimized searches for local alignments using a substitution matrix. However, as noted above, it would take a substantial amount of time to apply this strategy exhaustively. To improve speed, the program uses the observed pattern of word hits to identify potential matches before attempting the more time-consuming optimized search. The trade-off between speed and sensitivity is controlled by the *ktup* parameter, which specifies the size of a word. Increasing the value of *ktup* decreases the number of background word hits (i.e., those that do not mark the position of an optimal alignment). This, in turn, decreases the amount of optimized searching required and improves overall search speed. The default *ktup* value for comparing proteins is 2, but, for finding very distant relationships, it is recommended that it be reduced to 1.

The FASTA program does not investigate every word hit encountered, but instead looks initially for segments containing several nearby hits. By using a heuristic method, these segments are assigned scores and the score of the best segment found appears in the output as the init1 score. Several segments may then be combined and a new initn score is calculated from the ensemble. Most potential matches are then further evaluated by performing a search for a gapped local alignment that is constrained to a diagonal band centered around the best initial segment. The score of this optimized alignment is shown in the output as the opt score. For those alignments finally reported (a user-specified number from the top of the hit list), a full Smith-Waterman alignment search (without the constraining band) is performed. It should be noted that only the single optimal alignment is produced for each database sequence. As pointed out above, meaningful alignments can be missed by this approach if the proteins contain multiple modules. Consequently, it is recommended that matching sequences be further analyzed with the lalign program.

Beginning with version 2.0, FASTA provides an estimate of the statistical significance of each alignment found. The program assumes an extreme value distribution for random scores but with the use of a rewritten form of the probability density function in which the expected score is a linear function of the natural log of the length of the database sequence. Simple linear regression can then be used to calculate a normalized *Z*-score for each alignment. Finally, an expectation *E* is calculated, which gives the expected number of random alignments with *Z*-scores greater than or equal to the value observed.

BLAST

The BLAST programs introduced a number of refinements to database searching that improved overall search speed and put database searching on a firm statistical foundation. One innovation introduced in BLAST is the idea of *neighbourhood words*. Instead of requiring words to match exactly, a word hit is achieved if the word taken from the subject sequence has a score of at least *T* when a comparison is made using a substitution matrix to the word from the query. This strategy allows the word size (*W*) to be kept high (for speed) without sacrificing sensitivity. Thus, *T* becomes the critical parameter determining speed and sensitivity and *W* is rarely varied. If the value of *T* is increased, the number of background word hits will go down and the program will run faster. Reducing *T* allows more distant relationships to be found.

The occurrence of a word hit is followed by an attempt to find a locally optimal alignment whose score is at least equal to a score cutoff *S*. This is accomplished by iteratively extending the alignment both to the left and to the right, with accumulation of incremental scores for matches, mismatches, and the introduction of gaps. In practice, it is more convenient to specify an expectation cutoff *E*, which the program internally converts to an appropriate value of *S* (which would depend on the search context). In regions where matching residues are scarce, the cumulative score will begin to drop. As the mismatch and gap penalties mount, it becomes less likely that the score will rebound and ultimately reach *S*. This observation provides the basis for an additional heuristic whereby the extension of a hit is terminated when the reduction in score (relative to the maximum value encountered) exceeds the score dropoff threshold *X*. Using smaller values of *X* improves performance by reducing the time spent on unpromising hit extensions, at the expense of occasionally missing some true alignments.

There are several variants of BLAST, each distinguished by the type of sequence (DNA or protein) of the query and database sequences. The BLASTP program compares a protein query to a protein database. The corresponding program for nucleotide sequences is BLASTN. If the sequence types differ, the DNAsequence can be translated by the program (in all six reading frames) and compared to the protein sequence. BLASTX compares a DNA query sequence to the protein database, which is useful for analyzing new sequence data and ESTs. For a protein query against a nucleotide database, use the TBLASTN program. This is useful for finding unannotated coding regions in database sequences. A final variant is used only in specialized situations but is mentioned here for the sake of completeness: TBLASTX takes DNA query and database sequences, translates them both, and compares them as protein sequences. This program is mainly useful for comparisons of ESTs, where it is suspected that sequences may have coding potential even though the exact coding region has not been determined.

All of these programs make use of sequence databases located on server machines, which obviates the need for any local database maintenance. Some protein and nucleotide sequences databases currently available from the NCBI for BLAST searching. For routine searches, the *nr* database provides comprehensive collections of both amino acid and nucleotide sequence data, with redundancy reduced by merging sequences that are completely identical. To examine all sequences submitted or updated within the last 30 days, a database called *month* is provided. Both *nr* and *month* are updated on a daily basis. Several other databases are useful in more specialized situations, such as comparing against the complete genomes of model organisms (*ecoli* or *yeast*), searching specific classes of sequences (*est* or *sts*), or testing for the presence of contaminating or otherwise problematic sequences (*vector*, *alu*, or *mito*).

Table 10.1. Protein sequence databases for use with BLAST

Database	*Description*
nr	Non-redundant merge of SWISS-PROT, PIR, PRF, and proteins derived from GenBank coding sequences and PDB atomic coördinates
month	Subset of *nr* which is new or modi.ed within the last 30 days
swissprot	The SWISS-PROT database
pdb	Amino acid sequences parsed from atomic coordinates of three-dimensional structures
ecoli	Complete set of proteins encoded by the *E. coli* genome
yeast	Complete set of proteins encoded by the *S. cerevisiae* genome
drosoph	Complete set of proteins encoded by the *D. melanogaster* genome

An example of a BLAST search will serve to introduce various elements of a search output. For example, the amino acid sequence of one of the Alzheimer's disease susceptibility proteins (conceptual translation of GenBank L43964) was used as the query in a TBLASTX

Table 10.2. Nucleotide sequence databases for use with BLAST

Database	*Description*
nr	Nonredundant GenBank, excluding the EST, STS, and GSS divisions
month	Subset of *nr*, which is new or modified within the last 30 days
est	GenBank EST division (expressed sequence tags)
sts	GenBank STS division (sequence tagged sites)
htgs	GenBank HTG division (high-throughput genomic sequences)
gss	GenBank GSS division (genome survey sequences)
ecoli	Complete genomic sequence of *E. coli*
yeast	Complete genomic sequence of *S. cerevisiae*
drosoph	Complete genomic sequence of *D. melanogaster*
mito	Complete genomic sequences of vertebrate mitochondria
alu	Collection of primate Alu repeat sequences
vector	Collection of popular cloning vectors

search of the *est* database. One goal of such a search would be to identify cDNA clones for potential homologs in model organisms, thereby opening the door for experimental studies that would not be practical in humans (the clones corresponding to EST sequences are readily available). Each of the EST sequences in the database is translated in all reading frames before they are compared against the Alzheimer's protein sequence. The first two columns give the identifiers and descriptions for each sequence having a significant match. Although the definitions are truncated in this overview, the figures shows that sequences from both mouse and *Drosophila* are represented. The next column gives the reading frame that produced the best alignment (although there may be hits to translations from other frames as well). The next three columns provide the score of the best alignment, the sum *P*-value, and the number of HSPs that were used in the *P*-value calculation. There are actually two alignments involved, and scores are provided for each. In each case, the conceptual translation of the EST is shown aligned with the query sequence. Identical amino acids are echoed to the text line in between the sequences, and plus (+) symbols are used to indicate nonidentical residues that have positive substitution scores (i.e., conservative substitutions). It is noteworthy that the two alignments arise from different reading frames and are adjacent to one another, as can be seen from the sequence coordinates. This pattern is indicative of a reading frame error in the EST sequence. When analyzing sequence single-pass data, it is extremely useful to have tools that are relatively error tolerant.

Database Searching Artifacts

A query sequence that contains repetitive elements is likely to produce many false and confounding database matches. One clue that this may be a problem is the finding of significant matches to repeat "warning sequences" that have been included in both GenBank and SWISS-PROT. These entries are consensus sequences (or translations thereof) for different subfamilies of human Alu repeats. However, with the large amount of human genomic sequence now present in the database, it is common to have many hits to individual repeats with scores greater than those for any consensus repeat. Consequently, hits to Alu-warning entries are less striking than when the database was smaller. Other indications of likely artifacts would be finding hits to many proteins that seem to have no functional relationship to one another or hits to genomic sequences from many different chromosomes. These patterns might also be seen if both query and database are contaminated with foreign sequences from the same source, for instance, cloning vectors.

Although it is always good practice to critically evaluate database search results and be suspicious of artifacts when the data don't make sense, a more proactive approach involves *masking* problematic sequences in the query before doing the search. The problem of repetitive elements is ably handled by the popular program RepeatMasker, which identifies, classifies, and masks several types of repetitive elements and simple repeats. A masking strategy strategy, which we will call "hard masking," is to replace subsequences with an ambiguity character ("N" for nucleotide sequences or "X" for proteins).

Alternatively, a "soft-masking" approach, in which the resides are instead converted to lowercase letters, may be used with certain search programs. Because ambiguous residues are treated as mismatches (even when aligned to themselves), hard-masking effectively prohibits the identified repeats from making a positive contribution to the alignment score. Although hard masking is excellent for avoiding false hits, the fact that even the true alignments may be altered can present problems, particularly when alignment scores and lengths are used classify alignments. The solution to this dilemma is to use soft masking. Recent versions of the BLAST programs have an option that ignores regions of the query sequence that are lowercase when constructing the word dictionary. However, an alignment that is initiated in unique sequence may be extended through a repeat and would have the same alignment score as it would with unmasked sequence. With RepeatMasker, the —xsmall command-line option may be used for soft masking.

Both proteins and nucleic acids contain regions of biased composition, which can lead to confusing database search results. These *low-complexity regions* (LCRs) range from the obvious homopolymeric runs and short-period repeats to the more subtle cases where one or a few different residues may be overrepresented. Alignment of LCR-containing sequences is problematic because they do not fit the model of residue-by-residue sequence conservation. In some cases, the functionally relevant attributes may be only the periodicity or composition and not any specific sequence. Furthermore, methods for assessing the statistical significance of alignments are based on certain notions of randomness which LCRs do not obey. Consequently, many false positives may be observed in the output of a database search with an LCR-containing query sequence because the significance of matches can be overestimated.

A program called seg has been developed to partition a protein sequence into segments of low and high compositional complexity. Using this program, it has been shown that more than half of the proteins in the database contain at least one LCR. The evolutionary, functional, and structural properties of LCRs are not well understood. Perhaps LCRs arise by such mechanisms as polymerase slippage, biased nucleotide substitution, or unequal crossing-over. In proteins, LCRs are likely to exist structurally as nonglobular regions. Regions that have been defined physicochemically as nonglobular are usually identified correctly using seg. In DNA, there are many classes of satellite and microsatellite sequences that consist of many copies of a simple repeat unit.

The protein product of the human homolog of the *Drosophila* achaete-scute gene provides a good example of an LCR-containing protein. When analyzed with seg, two regions of low compositional complexity were identified. The first region is a 61-residue segment containing homopolymeric tracts of glutamine and alanine. The second is a 14-residue segment with a bias toward arginine. Without filtering, many database sequences with biased regions involving these amino acids would be reported. Using a command-line option, seg can generate the masked version of the sequence for use as a search query. Alternatively, filtering can be performed automatically by the BLAST programs through the use of optional parameters. Note that, in some implementations of BLAST, such as the Web version, filtering may be enabled by default.

Position-specific Scoring Matrices

In a standard substitution matrix, such as BLOSUM62, the substitution of one amino acid with another is associated with a single score—an obvious simplification given that the same amino acid may have different conservation patterns in one context than another in accordance with differing roles in biological function. Database searches can be tailored to find specific proteins families or domains through the use of substitution scores that reflect the substitution frequencies of each individual amino acid position in a domain.

There is a large literature on the construction and application of these *position-specific scoring matrices* (PSSMs), which may also be called hidden Markov models (HMMs), motifs, or profiles. In its simplest form, a PSSM consists of a set of 20 substitution scores at each position along the motif—one for each of the amino acids. Amino acids that are commonly found at a particular position receive higher scores, whereas lower scores correspond to amino acids unlikely to appear at that position. It is also possible to assign scores to insertions and deletions in a position-specific manner.

A commonly used software package, HMMER, contains a set of related programs for constructing and using PSSMs. Given a multiple alignment of several related proteins (e.g., one made using CLUSTAL W), the hmmbuild program may be used to calculate the position-specific scores and save it to a file (HMM file format). Using the hmmsearch program, the HMM file may be used as a query against a sequence database. Conversely, hmmpfam is used to compare a single query sequence against a database of PSSMs (HMMs). A comprehensive database of protein domains, Pfam, is often used for this purpose.

The power of PSSMs in database searches can be further enhanced by iterative approaches in which the highest scoring matches in one search are incorporated into a PSSM used in successive searches. Position-Specific Iterated BLAST (PSI-BLAST) provides an automated facility for constructing, refining, and searching PSSMs within the context of a single program. Starting with a query sequence provided by the user, the process begins with a standard BLASTP search of a sequence database. Highly significant alignments found in this search are then used to construct a PSSM on-the-fly. Comparisons of the PSSM against the sequence database are performed using a variation of the word-based BLAST algorithm used for standard sequence comparisons. The process continues until no new matches are found or a specified limit on number of iterations is reached.

To demonstrate the improved sensitivity of the PSI-BLAST approach, the sequence of histidine triad (HIT) protein was used as a database search query. Similarity between HIT and galactose-1-phosphate uridylyltransferase (GalT) has recently been described based on superimposition of their three-dimensional structures. However, sequence similarity between these two proteins is extremely weak. With a standard (single-pass) BLASTP search, no significant hits to GalT sequences are observed. The rat GalT protein is found in the second iteration and, after information from this alignment is incorporated into the profile, several additional homologs from other organisms are also identified.

Spliced Alignments

The identification of genes within long stretches of DNA sequence is a central problem for automatic annotation of complete genomes. For the very compact genomes of viruses and bacteria, this work amounts to little more than the enumeration of open reading frames. However, gene identification in eukaryotic genomes is significantly more challenging because of the larger amount of intergenic sequence and the fact that protein-coding regions may be interrupted by introns. The two fundamental strategies seek to identify genes using either the intrinsic signals in the DNA sequence or alignments to mRNA and protein sequences.

At first glance, the problem of aligning mRNA and genomic sequences seems trivial—using a local alignment strategy each exon would come out as a separate locally optimal alignment separated by large 'deletions' in the mRNA sequence corresponding to the introns that have been spliced out. If the goal is merely to obtain a crude sense of how a gene is organized, a simple alignment is sufficient. However, for the purpose of genome annotation, it is important that all exons be found with precise endpoints or a correct protein translation cannot be obtained. The sim4 program is designed to address genome annotation needs by performing mRNA/genomic alignments rapidly and accurately. It begins with a BLAST-like search for finding the obvious exons—those with very high alignment scores—and follows this with search at lower stringency to identify any missed (usually short) exons. Splice donor and acceptor signals in the genomic sequences are used to adjust the exon boundaries. To avoid problems caused by tandemly repeated genes, an additional constraint is imposed to require that the order of the exons found in the genome match that implied by the mRNA. Other programs available for performing mRNA/genomic alignments are est genome and the est2gen program from the Wise package.

It should be noted that an mRNA sequence may align perfectly well to a pseudogene and that such an alignment may be difficult to distinguish from a functional gene. Certain features may be indicative of retropseudogenes, that is those resulting from integration of a reverse-transcribed mRNA. For example, a poly(A) tract found in the genomic sequence at the 3' end of the gene is indicative of a pseudogene produced through an mRNA intermediate. Such an mRNA alignment will also lack the large gaps corresponding to introns, although this alone cannot be used to conclude that it is a retropseudogene because there many authentic genes that consist of a single exon. However, once all mRNA alignments have been generated for a genome, one strategy for retropseudogene identification involves looking for pairs of highly similar alignments in which one lacks and the other contains introns. Other types of pseudogenes may arise by gene duplication events followed by inactivation of one of the copies.

Such cases can be very difficult to diagnose. Programs that perform mRNA/genomic alignments usually do not use any knowledge of reading frame; therefore, determining that a potential gene contains a frameshift will require subsequent analysis of the protein translations. In the case of genes predicted by alignment with ESTs only, the protein-coding sequence is not known so any interruptions of the reading frame will not be apparent. The possibility that an apparent frame shift may actually be a sequencing error in the genomic sequence should also be considered, particularly in the analysis of a working draft sequence. In many cases, determining whether an mRNA-predicted gene is a functional gene or a pseudogene must await experimental validation.

Conclusions

Sequence alignment and database searching are performed tens of thousands of times per day by scientists around the world and represent critical techniques that all molecular biologists should be familiar with.

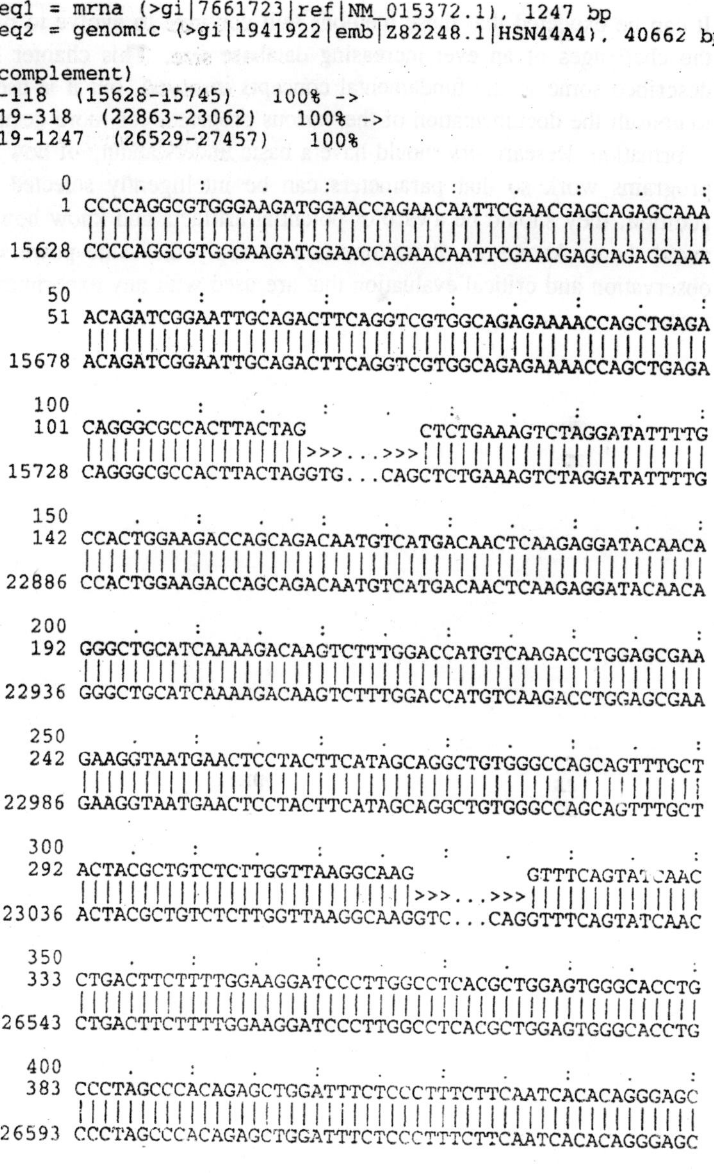

```
seq1 = mrna (>gi|7661723|ref|NM_015372.1), 1247 bp
seq2 = genomic (>gi|1941922|emb|Z82248.1|HSN44A4), 40662 bp
(complement)
1-118    (15628-15745)    100% ->
119-318   (22863-23062)    100% ->
319-1247  (26529-27457)    100%

      0     .    :    .    :    .    :    .    :    .    :
      1 CCCCAGGCGTGGGAAGATGGAACCAGAACAATTCGAACGAGCAGAGCAAA
        ||||||||||||||||||||||||||||||||||||||||||||||||||
  15628 CCCCAGGCGTGGGAAGATGGAACCAGAACAATTCGAACGAGCAGAGCAAA

     50     .    :    .    :    .    :    .    :    .    :
     51 ACAGATCGGAATTGCAGACTTCAGGTCGTGGCAGAGAAAACCAGCTGAGA
        ||||||||||||||||||||||||||||||||||||||||||||||||||
  15678 ACAGATCGGAATTGCAGACTTCAGGTCGTGGCAGAGAAAACCAGCTGAGA

    100     .    :    .    :    .    :    .    :    .    :
    101 CAGGGCGCCACTTACTAG         CTCTGAAAGTCTAGGATATTTTG
        ||||||||||||||||||>>>...>>>|||||||||||||||||||||||
  15728 CAGGGCGCCACTTACTAGGTG...CAGCTCTGAAAGTCTAGGATATTTTG

    150     .    :    .    :    .    :    .    :    .    :
    142 CCACTGGAAGACCAGCAGACAATGTCATGACAACTCAAGAGGATACAACA
        ||||||||||||||||||||||||||||||||||||||||||||||||||
  22886 CCACTGGAAGACCAGCAGACAATGTCATGACAACTCAAGAGGATACAACA

    200     .    :    .    :    .    :    .    :    .    :
    192 GGGCTGCATCAAAAGACAAGTCTTTGGACCATGTCAAGACCTGGAGCGAA
        ||||||||||||||||||||||||||||||||||||||||||||||||||
  22936 GGGCTGCATCAAAAGACAAGTCTTTGGACCATGTCAAGACCTGGAGCGAA

    250     .    :    .    :    .    :    .    :    .    :
    242 GAAGGTAATGAACTCCTACTTCATAGCAGGCTGTGGGCCAGCAGTTTGCT
        ||||||||||||||||||||||||||||||||||||||||||||||||||
  22986 GAAGGTAATGAACTCCTACTTCATAGCAGGCTGTGGGCCAGCAGTTTGCT

    300     .    :    .    :    .    :    .    :    .    :
    292 ACTACGCTGTCTCTTGGTTAAGGCAAG         GTTTCAGTATCAAC
        |||||||||||||||||||||||||||>>>...>>>||||||||||||||
  23036 ACTACGCTGTCTCTTGGTTAAGGCAAGGTC...CAGGTTTCAGTATCAAC

    350     .    :    .    :    .    :    .    :    .    :
    333 CTGACTTCTTTTGGAAGGATCCCTTGGCCTCACGCTGGAGTGGGCACCTG
        ||||||||||||||||||||||||||||||||||||||||||||||||||
  26543 CTGACTTCTTTTGGAAGGATCCCTTGGCCTCACGCTGGAGTGGGCACCTG

    400     .    :    .    :    .    :    .    :    .    :
    383 CCCTAGCCCACAGAGCTGGATTTCTCCCTTTCTTCAATCACACAGGGAGC
        ||||||||||||||||||||||||||||||||||||||||||||||||||
  26593 CCCTAGCCCACAGAGCTGGATTTCTCCCTTTCTTCAATCACACAGGGAGC
```

ouput truncated for brevity

Fig. 10.7. Spliced alignment. The sim4 program was used to align a novel human mRNA to the genomic sequence of a cosmid from chromosome 22 (EMBL Z82248). Three exons were identified on the complementary strand.

It can be expected that these methods will continue to evolve to meet the challenges of an ever-increasing database size. This chapter has described some of the fundamental concepts involved, but it is useful to consult the documentation of the various programs for more detailed information. Researchers should have a basic understanding of how the programs work so that parameters can be intelligently selected. In addition, they should be aware of potential artifacts and know how to avoid them. Above all, it is important to apply the same powers of observation and critical evaluation that are used with any experimental method.

INDEX